THRESHOLD

Rhetoric + Digitality

THRESHOLD

How Smart Homes Change Us Inside and Out

HEATHER SUZANNE WOODS

The University of Alabama Press
Tuscaloosa

The University of Alabama Press
Tuscaloosa, Alabama 35487-0380
uapress.ua.edu

Typeface: Minion Pro

Cover image: Illustration by GarkushaArt—stock.adobe.com

Cover design: Sandy Turner Jr.

Cataloging-in-Publication data is available from the Library of Congress.
ISBN: 978-0-8173-2194-9 (cloth)
ISBN: 978-0-8173-6143-3 (paper)
E-ISBN: 978-0-8173-9497-4

For Alex, home personified.

Contents

Illustrations

Figures

Tables

Preface

STAYING SAFE, STAYING SMART

On the first day of December in 2020, *The Atlantic* published an article with its thesis as title. The essay, entitled "Go Home Now," was a fervent, frazzled request, a demand, and a prayer. "A vaccine is waiting on the other side of this pandemic winter," author Derek Thompson wrote. "But before the temperature rises, and before the vials are distributed, Americans will have to find ways to remain safe inside, at home."[1] This article, along with many others like it, advocated a domestic strategy to contain COVID-19, the novel coronavirus that overwhelmed global populations starting in 2019. The argument of the essay was clear: getting its citizens to stay home was the best possible shot the United States had to stem the tide of COVID-19 infections. In popular and political pandemic discourse, home emerged as perhaps *the* single most important institution for fighting COVID-19 across the United States.

But not just any home. To manage the considerable demands of public life from the domestic sphere, the pandemic home had to be fortified with domestic technology that could connect the public and private.[2] For instance, communication conferencing platforms like Zoom enable some people to maintain work and school obligations from home. Contract tracing apps on smartphones encourage users to document and share their location when they leave the home. Devices with thermal sensors and cameras evaluate individuals' temperatures, alerting them (and possibly their employers) if they fall outside of normal ranges. Data collected by these smart devices can be activated to encourage or mandate particular actions. When paired with a networked surveillance infrastructure, "technology is being used to enforce stay-at-home mandates . . . by allowing authorities access to [individuals'] movement patterns."[3] Emerging technologies have long influenced the domestic experience, including in years prior to the pandemic. Yet during COVID-19 technology's reach into the domestic sphere expanded even further.

In other words, during COVID-19 many homes became *smarter*. This might sound like a good thing. But what "smart" means is not always clear, and the effects of embedding smart technology are not always immediately apparent. Introducing or proliferating smart technologies in the home has significant consequences for the privacy, health and safety, and agency of users and citizens writ large. So while some "are able to stay at home because of digital technologies and the data they produce . . . the same technologies that enable our lockdown lives now could normalize the heightened surveillance and control creep post-pandemic."[4] Importantly, the mediating effects of smart technology don't stop at the front door of the smart home. Rather, smart technologies are remaking home for all of us—whether or not we use the technologies, whether or not we are at home.

MAKING A HOME SMART

This is a book about smart homes: what they are, what makes them smart, what we can expect from them, who benefits from them. Generally speaking, smart homes are houses that are equipped with networked technologies. Frequently, they deploy "smart devices": domestic tools, appliances, and gadgets that connect to the internet and/or other smart devices. Smart homes can be built smart. Other homes are made smart by virtue of the incorporation of smart devices. Still others are smart because of a spatial intelligence that supports the needs and desires of people inside and outside of the home.

What makes a home "smart" is not only its technological capacities but also its relevance to society. What is smart about smart homes is therefore profoundly contextual, such that something that was smart in one era may be obsolete by the next one. And yet the smart home as we understand it today is inextricably connected with domestic technologies of the past. For example, the idea that the collective pandemic home ought to be a smart, hermetically sealed, and protected space is not unique to the COVID-19 era.[5] Rather, the COVID-19 era smart home is the product of previous domestic builds and sensibilities.

During the eighteenth and nineteenth centuries, bourgeois conceptions of the domestic ideal were formulated even as new comforts were made more readily available through technological innovation.[6] As a force for moral goodness, tranquility, and status, bourgeois domesticity was architected in part by delineating inside from outside.[7] The respectable domestic interior was deemed worthy of protection; the amorphously defined exterior served as the necessary corollary to construct and qualify threats to the interior.[8] Relatedly, in the context of the United States of America, the idealized home served as a site to expand the reach and scope of the normative American way of life.[9] Citizenship could be practiced from the domestic space, even as popular discourse placed the home in opposition to the public world of politics. While the bounds of the division between public and private would shift over time, the division itself would persist, calcify, and even expand outward beyond the home. In a theme that would repeat in the smart home, technological systems built to manage the domestic sphere intersected the domestic ideal, helping to shape the configuration of homes and the uneven social relations that occurred inside and outside of them.[10]

In the twentieth century, the domestic ideal of the nineteenth century expanded and contracted to negotiate social and technological upheaval, which often went hand in hand in influencing housing policies and practices. In this century the home remained a status symbol. Its physical and symbolic boundaries were enforced through various social and technical mechanisms, with knock-on effects felt even today. In the first third of the twentieth century, for instance, New Deal legislation increased white families' chances of homeownership while institutionally encouraging segregation in mortgage lending—a practice that would be replicated later in digital technologies.[11] Model homes were built to showcase "the house of tomorrow"; they featured new spatialized technologies like centralized kitchens and remote garage door openers that helped further shape a bourgeois domestic technological ideal in the United States.[12] Sarah J. Darby has argued that the "concept of a smart home can be traced back to [these] futuristic display homes" but that the "concept remained a specialized one for some time, able to take shape for a mass market" only much later in the twentieth century.[13] Ahead of this expansion, domestic technologies like television shaped domestic space and social relations in new ways, reprising Victorian domesticity through "a suburban consumer culture."[14] US populations used their televisions to watch the Moon walk, the war in Vietnam, and marches for civil rights, among other high-profile events, further merging the domestic interior and exterior. At the same time, the television afforded nuclear families the capacity to remain firmly inside the (increasingly suburban) home while seeing the world change outside of it. As white men prepared to walk on the moon, cultural leaders decried the continuing concentration of the riches of technological progress in the hands of the few.[15] While many historical happenings occurred on a cosmic or existential scale, popular discourses at the time were quick to link them to domestic pursuits like housing inequality.[16]

After the upheaval of the middle century, the 1980s seemed to usher in a new era of American traditionalism that would influence the formation of the smart home as we know it today. The era was characterized by a number of sociopolitical forces bringing about emerging forms of neoconservatism.[17] Mass movements from the middle century—civil rights, women's rights—were similarly rejected by institutions and people in power. The power of the collective was strategically rerouted with increased focus on the protection of the rights of the individual. Neoconservatism helped to reaffirm the domestic sphere as a private space for a particular type of family, rework nostalgic forms of domestic subjectivity introduced a century earlier, and virulently defend so-called free market economics.[18] It is perhaps unsurprising, then, that the smart home *began* as a marketing technique connecting informational technology to a privatized domestic sphere. The phrase "smart home" was first used in the mid-1980s by a residential builder association in United States.[19] The phrase "captured some of the meaning of the older term 'wired homes' . . . while riding on the proliferating use of 'smart' as a popular way of branding new products exploring the potential of information and computer technologies."[20] The smart home was positioned not only as an architectural ideal for building the next phase of the American home but also as a framework for selling it to wealthier classes of American people. Domestic technology played three fundamentally conservative roles: to maintain what had come before under the signifier of futuristic technologies; to bound the interior while linking it to the outside world through information technologies; and to craft an ideal domestic subject whose lives would be organized through increasingly proliferating computational technologies.

Now we tend to think of smart homes as those outfitted in digital ("smart") technologies. The smartest homes appear to be those with sleek appliances that automate chores, with chummy virtual assistants that tell jokes and turn off the lights at night, and with spaces that gleam with the glitz and glam of a science-fiction aesthetic. This book certainly examines these types of smart homes. But it also explores other types of smart homes, like "ordinary" homes that quietly incorporate smart technologies, linking the domestic space to the wide world of the platform without much fanfare. It considers smart homes that aren't "born" smart. These homes are made smart by deliberate integration of off-the-shelf technologies, retrofitting not-smart homes to smart homes incrementally. Differing configurations of the smart home provide evidence that the smart home is both a spatial arrangement and a set of meanings that cohere—or don't—within broader cultural narratives about both home and technology.

PANDEMIC DOMESTICITY: A RENEWED COMMITMENT TO HOME

The register of "smart" occupies a complex constellation of meanings that correspond closely (but not always perfectly) to the sociotechnical moment in which they are made, used, and given meaning. This, then, is a book about technology, but it's also necessarily a book about how the social influences the creation and use of technology in domestic spaces. COVID-19 has fundamentally altered the domestic narrative, drawing out into the open all the reasons why "home" and the related concept of domesticity remain crucially important for a functioning, healthy, and egalitarian society. But COVID-19 has also made it abundantly clear that, smart or not, home is not an institution accessible to everyone (or even to most people). As more homes become smart, we have an opportunity to evaluate the future they offer. That is what I hope to do in this book.

Before we begin, a bit of background is necessary. When I started to write this book in 2018, smart technology was transitioning from its early-adopter roots to a fairly standard household feature. Introduced a few years earlier, Amazon's Echo and Google's Home Assistant devices were starting to gain traction as household names. In the midst of this transition, I was interested to know how domestic technologies might influence how we understood "home." I believed then, as I do now, that smart technologies were (and are) rewriting domestic scripts about what it means to be at home. But it wasn't until COVID-19 that I fully understood that the smart home is one point of arrival in a long lineage of domestic technologies, and that, like other histories, the history of technology and domesticity repeats itself in ways both predictable and unpredictable.

As a scholar of rhetoric and technology, I am interested in charting the communicative processes that made sense of the smart home through these changes. I wanted to study how people designing, making, selling, and using smart technology described the smart home and gave language to the impacts it has on domestic life. Furthermore, I was drawn to the smart home as a site for rhetorical research because it is both aspirational and practical—it is a space used by people to imagine future possibilities *and* a place where everyday life is lived. The ultimate promise of smart technologies is that the smart home can bridge the two. That is, if done correctly, the smart home should elevate us all to a technological ideal through the mediation of ordinary actions. In the quiet moments of the everyday, the pitch goes, smart technologies can quietly optimize our lives—even when they're threatened by a virus.

By 2020 smart technology had a new job for which it was perfectly (if not initially) suited. When governments announced stay-at-home orders, smart technology was pressed into service to help manage effects associated with the pandemic. As roads, theaters, shopping malls, schools, and universities emptied out, many—but not all—workplaces shifted to remote. Schools turned into online learning environments. As a result, during the multiyear pandemic, millions of people in the United States of America found themselves at home for longer periods than ever before. Even later in the pandemic, when travel and other restrictions loosened, the home remained an important site for transmission mitigation for sick (or suspected sick) people. The domestic sphere shifted to accommodate these changes spatially and figuratively, as it has done many times before.[21]

In the domestic interior, smart technology could be used to connect with others across pandemic bubbles. Platforms sprang up to encourage remote socialization through media consumption; spare rooms or corners in houses or apartments became Zoom rooms. Some people leaned into life at home by reengaging with domestic arts through networked technologies. There was a run on flour as residents learned how to make sourdough starters and bake bread with online tutorials. Google Trends reported that people searching for "recipe" reached an "all time high."[22] Cleaning regimens populated general media outlets in addition to domestic-minded websites like Apartment Therapy. A number of digital outlets published guides on how to homeschool children, how to entertain them, and how to keep working when children were not in school or daycare. Other outlets encouraged us to sit with the stillness, and to care for ourselves by caring for others. In providing care for others, some domestic subjects privately performed an ethic of community from the domestic space. Some even began "Victory Gardens," harking back to a collective memory of when the nation was at war and citizens were expected to engage in collective efforts as part of the American project.[23]

And yet, even as the home potentially serves as a site for collective politics, it has been repeatedly rendered a site for individual protection. Outside the domestic interior, smart technology managed workers, supply chains, and logistics processes—sometimes all at once. Grocery apps employed predictive algorithms and other computational logics to regulate supply and demand of food, household goods, and the labor of grocery store workers. Networked surveillance applications monitored the temperature of workers in warehouses as they "picked" and packed boxes to deliver goods to households. As I wrote the prose that became the book you are reading now, I watched in real time as a familiar hierarchy formed relative to the domestic space. As before, one's status in the domestic pandemic hierarchy depended a great deal on whether one was figured as domestic insider or outsider—whether the smart home products were supposed to work for you or to make you work (harder, faster, for less money). Although this particular social arrangement was mediated through new technologies, the home as a social space for determining insider versus outsider persists throughout time. COVID-19's reworking of the connected home echoes important historical and social phenomena regarding domesticity as a site managing the boundaries of self and other, of public and private. Historically, homes link the interiority of individualism with the collectivity of the social, political, and economic spheres. Smart homes, in turn, link the interior of the domestic sphere with the external world through platforms, networks, code, and other sociotechnical logics.

Conditions outside of the home continue to influence the world inside of it. During the pandemic, local and national economies shuttered. In March of 2020, the Dow Jones industrial average fell almost three thousand points, "its worst day since the 'Black Monday' market crash in 1987 and its third-worst day ever."[24] The NASDAQ hit the lowest point ever recorded.[25] Oil futures crashed, dipping below $0 for the first time ever.[26] Gas prices dropped precipitously, as low as 89 cents in some parts of the country.[27] Unemployment sharply increased, with "more than 33 million unemployed, or a real unemployment rate of 20.6%—which would be the highest level since 1934."[28] Local independent business shut down, leaving thousands jobless. What remained of local economies was propped up by essential workers, many of whom were working for minimum wage at big chains. Median household income in the United States decreased nearly 3 percent in 2020, "the first statistically significant decline in median household income since 2011."[29] A 2022 report by the Brookings Institution describes "an increase in 'extreme poverty' by 115 million people in 2021 because of the COVID-19 effects."[30] As I revise this book, financial experts warn of another global recession. This time, gas prices are rising, as is inflation.

As the poor get poorer, the rich continue to get richer. Billions of dollars richer. *USA Today* reported that while "the COVID-19 pandemic has triggered an economic crisis of a magnitude not seen since the Great Depression," in which "more than 11 million Americans remain unemployed and many shops and restaurants will never reopen," US billionaires "grew their net worth [by] 931 billion."[31] Smart homes play a noticeable role in propping up this lopsided economy. The so-called platform economy, which is the backbone of many smart home services, generates immense wealth for corporate oligarchs while keeping contingent workers working harder for less. Jeff Bezos, founder and former CEO of Amazon, a leader in the smart home space, increased his net worth by over $90 billion during the pandemic.[32]

The pandemic continues to ravage the housing sector. Housing prices are at historic highs, pricing even well-off buyers and renters out of the market. As the economy slumped and unemployment skyrocketed, many Americans lost the ability to pay for housing. Some governmental mandates attempted to ease the housing crisis—for a limited amount of time. For instance, agency orders by the Centers for Disease Control and Prevention (CDC) and the Department of Health and Human Services (HHS) provided relief by placing a moratorium on evictions for those who qualified. Those orders did not forgive rent or mortgage payments, pushing many into further debt as the pandemic progressed. The eviction moratorium announced by the CDC and HHS also did not "preclude the charging or collecting of fees, penalties, or interest as a result of the failure to pay rent or other housing payment on a timely basis."[33] For those without employment (or with low-wage jobs), relying on an overstretched social safety net, such measures meant they were facing the prospect of both a deadly disease and significant debt. There are reports that landlords filed thousands of evictions, putting Americans at risk for homelessness *and*

COVID-19. This is doubly concerning, as being unhoused is also an increased risk factor for COVID-19.[34]

As we collect more data, we are learning that COVID-19 did not kill indiscriminately. Study after study shows that accounts of COVID-19 as the "great equalizer"—"implicitly arguing that the virus has affected everyone regardless of race, ethnicity, sexuality, income, or zip-code—are wrong."[35] In the United States, the populations hardest hit by the pandemic tended to be the most socially and politically vulnerable populations, including poor and working-class people, individuals with underlying medical conditions, racial and ethnic minorities, disabled people, and the homeless. Researchers examining health disparities in COVID-19 transmission and treatment noted that "for historically disadvantaged populations, who experience fractured access to healthcare under standard conditions and who are more dependent on low-wage or hourly paid employment, the pandemic has had a disproportionate impact."[36] For these populations, staying home to flatten the curve was not a realistic choice to be made. Still, these populations were subject to the expansive reach of digital technologies that mediated their actions, movement, and existence according to amorphous frameworks of "smartness."

A NEW SMART HOME FUTURE

COVID-19 remains a truly global crisis, in which half a billion people have contracted the disease and millions of people have died.[37] Although I began writing this book in my campus office, the pandemic quickly relegated me to my home to finish it. As I watched the coronavirus spread across the United States, including to my home state of Kansas, the relevance of domesticity sharpened in my mind. COVID-19 has shown that access to safe, secure housing remains the foundation for a healthy, productive, and just society. Paradoxically, the pandemic has also created new opportunities for bourgeois domesticity to be used as a tool for concentrating power and wealth. The current smart home continues on in this trajectory, intensifying the division of domestic classes with the mediating force of networked technologies. Fear and uncertainty related to COVID-19 offer further excuse to maintain (and even intensify) social and economic division.

From my perspective, the smartest home is not the most digitally connected home. Rather, it is the home that keeps people safe, connects them to their community, and thoughtfully integrates domestic technologies to enable people to live a life full of dignity, care, and good health. A truly meaningful approach to the smart home must expand the definition of "smart": "smart" ought to mean accessible; ought to mean clever and thoughtful in response to a collective social challenge; ought to mean caring for the collective; ought to mean networked (as in community). The signifier of "smart" should clarify whether or not we are using all available tools to extend that safety, health, and dignity to people beyond our own households. This type of smart home can serve as a life raft for weathering crises, caring for one another, and engaging in new forms of networked sociality. This is the smart home future I hope for. We must build it together.

Acknowledgments

WRITING A BOOK about the smart home during a pandemic provides a unique kind of clarity about the value of one's intellectual community. Even though we were apart for much of the time I was drafting this manuscript, I am indebted to an exceptional group of people who made this work possible.

This book benefited from sharp graduate research assistants at Kansas State University. Marlene Pierce did exceptional audio transcription work of fieldwork that became parts of chapters 1 and 2. Lora Kirmer assisted with research on the history of the smart home and coauthored a related essay on the Amazon Echo Auto that was presented at the 2020 National Communication Association annual conference. Ariel Lambert assisted with early copyediting and materials acquisition, particularly for chapter 2. Nicholas Aranda provided astute editorial insight on later drafts. In addition to offering copyedits and supporting formatting efforts, Nicholas also offered thoughtful insight and criticism on the theory of living in digitality. More generally, my undergraduate and graduate seminars at Kansas State are ideal intellectual communities, creating spaces to generate and work through ideas related to rhetoric, space, and emerging media. My graduate seminar in communication and technology deserves particular recognition for offering useful feedback on the lectures that would become sections of this book. To my students: I am grateful for your smart contributions.

My writing group, composed of Caitlin Bruce, Michaela Frischherz, Lisa Silvestri, Kim Singletary, and Emily Winderman, reviewed numerous early drafts of various chapters of this book. In addition to criticism, they provided care and camaraderie—a real homeplace in the discipline. Long live the LDP.

Atilla Hallsby and Nikki Marcotte read early (and late) chapter drafts and also supported the vision of this book with collaborative brainstorming and resource sharing. I'm also grateful for my pandemic Zoom "body double" writing crew, including Elizabeth Mendenhall, Natalie Pennington, and Dustin Greenwalt. I could not have completed the book without you. Grace Duque and Stephen Loch challenged me to consider radical possibilities within a community of care. Thanks, you two. Leslie Hahner provided characteristically exceptional advice and the infamous "Hahner edit."

Much gratitude to Dan Waterman at the University of Alabama Press. Dan's generosity of time and expertise has made the submission and editing process joyful. Two anonymous reviewers offered valuable feedback that sharpened the focus and prose of this book—and they did it in the middle of a global pandemic. Thank you for encouraging me to find and occupy the true home of the manuscript. The coeditors of the Rhetoric + Digitality series, Casey Boyle, Michelle Kennerly, and Damien Smith Pfister, are an author's dream (team). They offered support, guidance, and generous, critical feedback that helped move this project forward.

Alex McVey is an extraordinary life partner, dog dad, and interlocutor. While I wrote the book, he lived (somewhat begrudgingly) with the smart technologies I installed throughout the house for research purposes. He also cooked, cleaned, and performed other integral domestic labor—much less begrudgingly. Thanks for helping me build a smart homeplace.

In a book about the spatialization of digitality, it makes sense to me to acknowledge the spaces around me that influenced the ideas that became the manuscript. Parts of this book were written on campus at Kansas State University, in two offices and a handful of seminar rooms in Nichols Hall. The community and quiet of the basement at Auntie Mae's in Manhattan, Kansas, supported the germination of ideas about finding home outside the home. I wrote other sections of the manuscript on tiny café tables in DFW and Chicago O'Hare while the frenetic movement of people around me reminded me of the scope of placemaking. Finally, when the pandemic ushered me home to write, my little midcentury modern home surrounded by wild flora and fauna and encouraging neighbors kept me tethered—mostly.

Threshold was supported by two grants from Kansas State University. The University Small Research Grant provided funds for fieldwork at the Imagine! Smart Home in Colorado and at Duke University in North Carolina, as well as research and technology support. The Faculty Enhancement Award supported additional fieldwork in San Francisco and the DC metro area, as well as a summer sabbatical. Thank you to Sarah Riforgiate, who served as my institutional mentor during these funding processes. My department chair also helped secure funds to support related research trips and equipment. Thanks, Greg Paul!

Thank you to the intellectual communities who invited me in while I was stitching together early ideas about the smart homeplace. An earlier version of the introduction was presented as an invited lecture at the University of Kansas in spring 2021. Chapter 1 of this book is derived in part from an article published in *Cultural Studies*, March 26, 2021, © Taylor & Francis, available online: http://www.tandfonline.com/10.1080/09502386.2021.1895254. I am grateful for Blake Hallinan and James N. Gilmore's editorial insight and encouragement.

Several individuals made my fieldwork possible. Alex McVey took the photos included in chapter 1. Thank you to Imagine! Colorado, Brookfield Residential, Duke University, and Parks Associates for inviting me into your smart home (conferences.) To the residents of the Duke University Home Depot Smart Home and the Charles Family Smart Home, extra-special thanks are in order. I appreciate your generosity and willingness to open your homes to me. Parks Associates also provided a complimentary pass to their 2020 CONNECTIONS smart home conference so that I could continue my research while working from home. Thank you!

THRESHOLD

Introduction

"Homes . . . are metamachines of literally thousands of different technological components."

—Kitchin and Dodge, *Code/Space*

"As we write, so we build: to keep a record of what matters to us."

—Alain de Botton, *The Architecture of Happiness*

"All my life I have searched for a place of belonging, a place that would become home. . . . Home was the place where the me of me mattered. Home was the place I longed for; it was not where I lived."

—bell hooks, *Belonging: A Culture of Place*

Our homes are increasingly "smart." Smart homes outfit domestic spaces with emerging technologies, including appliances, gadgets, platforms, networks, sensors, etc. For example, smart homes can customize a user's domestic experience by creating ambiance with Wi-Fi-connected speakers and lightbulbs, providing soft lighting and soft music every day at dinnertime. Smart homes purport to save energy and create greater feelings of comfort by raising and lowering a platform-connected thermostat at night and when homeowners are away. Smart homes use automated technologies like smart vacuums, ovens, and fridges to create pockets of leisure time for homeowners as smart appliances monitor groceries, cook food, or clean the kitchen floors. Smart homes promise to keep people in touch across distances through devices with embedded sensors, cameras, speakers, and other gadgets. While users sleep, intelligent technologies can monitor the chemical makeup of the atmosphere in the home, trigger emergency lights, and alert users to movement outside the home. Even the bed may be smart, tracking the movements of sleeping bodies at rest. Smart homes immerse their users in networked technologies from the time they rise in the morning to the time they sleep at night and all the time between.

Industry experts anticipate that soon nearly half of all American homes will be smart homes.[1] In response, home builders are expanding their lineup of tech-enabled houses. These smart-from-the-start houses—many of which appear quite traditional on the outside—are remarkable in that they build in smart technology to create a souped-up, high-tech version of the US dream home. These smart homes feature Wi-Fi-connected appliances, bespoke surveillance systems, and even new rooms dedicated to remotely receiving goods purchased through Instacart or Amazon.

At the same time, companies that create smart devices compete fiercely for space in existing American homes. Indeed, a sizable portion of the smart home market seeks to bootstrap the modern American house to its most intelligent (or connected) form. Smart speakers, Wi-Fi-enabled plugs, and plug-and-play virtual assistants are designed to appeal to the millions of Americans whose houses are not yet smart—but could be. The smart home is a place and an aspiration: domestic environments that aren't yet smart might become so, with just a few strategic purchases or installations.

The smart home market is lucrative. Many of the world's wealthiest and most influential companies have entered the smart home space, producing goods and offering services to aid in the smart home transformation. These companies include the usual suspects, like Google, Facebook, Amazon, and Apple. However, the Silicon Valley elite are not the only important actors in smart home development. Appliance makers, logistics companies, health and wellness powerhouses, insurance outfits, and security franchises are all betting on the smart home in an economy that puts a premium on data. Together, major players in the smart home space have successfully ensnared the attention and pocketbooks of millions of households by touting the virtues of ambient, networked technologies as an upgrade to modern domestic life.

Many homes across America are already smart. More are on the way to becoming smart. As of 2019 consumers have purchased a reported 100 million devices configured with Amazon's virtual assistant, Alexa.[2] Google boasts that its smart assistant "is now available in more than 400 million devices" in "a dozen countries . . . offering help in 8 languages around the globe."[3] In 2017 alone smart home consumers purchased more than six million smart speakers from Google. Other companies seek to expand the reach of the smart home not through new devices but by revitalizing technology already in the home. Samsung, a leading purveyor of smart appliances, launched an initiative to turn old smartphones into smart devices for the home. By repurposing ubiquitous technologies already owned by consumers, "devices currently not in use can play an important role in turning every home into a smart home."[4] The smart home vision joins connected technology with domestic space and, in so doing, changes the way we live life at home. The current telos of the smart home is characterized by technology's complete and total incorporation in the domestic sphere.

Current events encourage the rapid adoption of smart technologies at home. I wrote this book in the middle of a global pandemic that saw (and *sees*) increased reliance on connected, digital technologies at home and in public. For instance, the pandemic accelerated the adoption of smart home technologies as governments mandated stay-at-home policies.[5] Homes were—and continue to be—utilized as recovery sites, remote offices, doctors' offices, nurseries, and classrooms. Any spare space might be conscripted to be a "Zoom room."[6] As the homebound reconfigured domestic space to meet the challenges of pandemic life, the COVID-19 pandemic helped normalize the use of broad, widely distributed surveillance measures to track the spread of the virus and the movement of various populations, "rescripting privacy in the process."[7] Global crises thus helped smart home technologies gain a foothold at home and in the cultural imaginary, particularly when the home serves as a key tool in responding to an epidemiological threat.

The rapid speed at which smart homes are built or made smart demands our immediate attention. As smart homes proliferate across the United States, they bring with them implicit expectations about the use of domestic technologies and configurations of domestic space in the twenty-first century. This book therefore explores these questions: How do smart homes communicate cultural values about the role of technology in the twenty-first century? How do these technological and cultural values, in turn, shift or augment domestic spaces? Who do these new smart homes serve, and who might they serve if we are intentional about their use? Answering these questions is time-sensitive, as the next few years will determine how smart homes are configured, who has access to them, and what they mean.

The primary argument of this book is that smart homes make a normative claim regarding the favorable and inevitable role of digital technologies in everyday life that translates spatially in the smart home. This normative claim turns on and intensifies already-existing cultural norms celebrating both extensive technological proliferation and digital capture of all parts of life. Moreover, smart home discourses interact with the home as a rich, meaningful site for living and engaging with others. This book shows how smart technologies provide a literal and figurative blueprint to configure the collective meaning of home in ways that challenge and reinscribe existing expectations for homelife. The book will also highlight how smart homes goad the spatial enactment of ubiquitous computing as a normalized, daily practice, readying domestic spaces and their occupants for an increasingly transactional digital future that is largely controlled by technological hegemons.

Smart homes are spatial manifestations of the sociotechnical imperative to engage with technology at every moment of every day. Their existence and continual proliferation direct users' attention and action within the domestic space and, more broadly, help shape the domestic ideal to be compatible with smart technology. As such, a secondary claim of this book is that smart products are culturally relevant insofar as they are *agential* and *transformative*: smart homes deploy intelligent or connected technologies to calibrate life for the digital age. Smart products—and those who make, sell, and use them—orient spaces according to technological logics. Sometimes technological proliferation manifestly reworks domestic spaces to appear positively space-age, tricked out with all the paraphernalia of futurism. In many other instances, smart technologies simmer under the surface of the modern home space, infusing an existing domestic ideal with new tools to bring the technological vision to life quietly. In these cases the smart home's futurism functions as a time machine back to an earlier, supposedly easier, domestic existence for an overwhelmed and overburdened domestic subject.

An important way smart home discourses manifest in the cultural imaginary is through futuristic, utopian rhetorics. Smart homes purport to enact a domestic utopia through the digital capture and subsequent reconfiguration of habits, routines, processes, and homelife activities. One expression of smart utopia manifests as techno-solutionism. From all-in-one cooking devices to automated security systems, smart products are marketed to consumers as miracle cures for the domestic trials endemic to the modern condition. In times of crisis, smart technologies are lauded as ways to right the ship quickly—never mind privacy or equity concerns. More generally, smart homes—and the people who make and sell them—help to construct a mythic ideal of a good life.[8] To be clear, the techno-optimist urge to upgrade life with novel technologies "resonates with and repackages technological utopian ideals from the past."[9] What's new is the intensity and rapidity with which smart technologies are integrated across every aspect of modern life. Smart devices are frequently embedded in non-smart objects such that a life caught up in the Internet of Things becomes virtually inescapable. In addition to prompting troubling questions regarding the agency or privacy of smart home users (and nonusers), the utopian rhetorics that correspond to smart technologies belie (or outright deny) the power structures underscoring the development and use of smart technologies.

Smart homes routinely re-create or affirm existing social hierarchies in the public and private spheres. Smart homes rely upon and intensify hidden social and technological systems, including biased algorithms, toxic ideologies, and more. Because smart tools are not neutral, their use in the domestic space has political, social, and economic effects that are unevenly distributed across populations. Examples of inequity in the smart home space are apparent if one only looks for them. Do-it-yourself or professionally installed smart security systems have been repeatedly used to racially profile people of color, fueling America's ongoing racial apartheid in and around the home. Smart homes are increasingly important tools in the ongoing process of gentrification. In some cases smart technology is used as a pull to encourage homebuyers to purchase homes in a newly "revitalized" residential area. As seen in chapter 1, the incorporation of smart-from-the-start homes into an already-existing community can drive up real estate prices, pricing long-term community members out of these spaces. In these communities, social and economic hierarchies are (re)produced or else reified by geographic and logistical changes, inviting homogenization and the concentration of wealth and power while alienating vulnerable populations. Smart homes' all-expansive techno-solutionism is especially pernicious as smart technology is routinely pitched to residents and city administrators as a high-tech amenity meant to serve an entire community.[10] From this perspective smart devices increase the efficiency of residential services like utilities or transit. Yet smart technologies cannot serve individuals who do not already have meaningful access to basic services or utilities in the first place.[11] In general, the logics of these smart systems are built to serve the needs and interests of a privileged class.

As it exists now, the discursive and spatial landscape of the smart home privileges homeowners and those inside of the household at the expense of those framed as outsiders. The home has historically been a site of unequal labor relations, and the introduction of smart technology does not, as some might claim, relieve these inequities. In fact, smart homes routinely exacerbate power differences between groups of people working inside and around the home. For instance, smart homes intentionally plug users into the platformed gig economy. This provides a leisurely shopping experience for customers by outsourcing labor to contingent workers whose minute-by-minute movements are collected, analyzed, and optimized. One result of this optimization is that Amazon delivery drivers "don't have the time" to use restroom facilities on the clock. To make the delivery timetables work, some are forced to urinate in water bottles.[12] The smart domestic space also amplifies and extends traditional patriarchal gender roles of the domestic sphere. Virtual assistants like Siri, Alexa, Cortana, and others were modeled on stereotypically feminine characters. They continue to perform gendered labor in the home, reprising the role of the ideal 1950s housewife.[13] In these ways smart homes structurally reify existing domestic hierarchies in the United States of America, producing a new set of valued (and valuable) digital subjects. At the same time, the modern smart home assumes and enables the formation of a set of digital others whose value is articulated through capitalist frameworks of surveillance, extraction, and technological management of their bodies and time.[14]

While smart homes may, at the outset, appear to signify a futuristic and therefore more equitable way of living, they transport and intensify deep-seated social hierarchies, stitching together complex histories of domestic inequality with digital space. The result is a more efficient, insidious, and powerful form of social control, where political and technological mediation is routine, complete, and often invisible. Invisibility gives smart homes additional power not only to direct our actions but to shape our worldview without us realizing it. Paul N. Edwards has remarked that "the most salient characteristic of technology in the modern (industrial and postindustrial) world is the degree to which most technology is *not* salient for most people, most of the time . . . mature technological systems . . . reside in a naturalized background, as ordinary and unremarkable as trees, daylight, and dirt."[15] In other words, humans are building a social and technological infrastructure that is totalizing and yet banal, remarkable in its unremarkability. The smart home is an example of how these powerful, connected, and distributed technological systems create a seamless digital environment in which to live with smart technologies. While these built structures currently serve a few classes of privileged digital subjects by imbuing their spaces with useful technological features, the final horizon of the smart home vision will place agency and control over homelife firmly in the hands of technological oligarchs who imagine a profitable domestic future to increase their power and position. Our job as critics, thinkers, advocates, and citizens is to intervene in these processes and architect more just digital futures in which to live.

In sum, smart homes and their advocates help articulate a spatial, sociotechnical vision for how life should be lived. This book examines this pro-technology mindset and how it has manifested in the smart domestic space. The book also provides evidence that absent intervention, the current trajectory of the smart home reinvests power in privileged domestic subjects who are already prioritized by the present digital order. Distributing the power of smart technologies to broader publics is necessary and will require reimagining both the home space and the function of technology in society. Rethinking inegalitarian use of smart technologies means first understanding how particular cultural imaginaries around space and smartness take hold.

LIVING IN DIGITALITY

Smart homes—and those who produce, market, and use them—don't just scaffold smart technologies onto already existing spaces. Smart homes *transform* space, staging the domestic sphere as a nodal point in a larger digital society. In so doing, smart homes invite users to engage with increasingly ubiquitous digital technologies as daily routine. Because smart homes represent an opportunity to have one's whole life mediated by digital technologies at all times, they create an opportunity to *live in digitality*. Living in digitality is both a cultural condition and a set of mediated practices in which the sociotechnical affordances of digitality are relayed to the domestic sphere. Here, technology is not just a set of discrete objects or artifacts with which to engage. Nor is technology simply a set of tools for solving a problem. Rather, living in digitality gestures toward the totalizing, immersive condition of living with smart products and the worldview they engender. Smart products are environmental; they manifest in space. More than that, smart technologies *articulate* a spatial ideal and, with the help of the humans who interact with them, produce it for privileged digital subjects. As a set of cultural practices, living in digitality describes the actions that exist along the intersection of the human and nonhuman. Living in digitality manifests spatially through these conscious and subconscious actions; smart spaces encourage domestic subjects to live every day in digitality. I spend the remainder of this text examining the spatial production of the smart home across multiple case studies. For now, I begin by characterizing living in digitality as a framework for living within and beyond the smart home.

There are seven key facets of living in digitality: optimization, scale, computation, generalizability, control, bias, and incorporation (see Table 1). Each of these components works with the others to support the smart home as a primarily spatial manifestation of digitality. At the outset it is important to clarify that living in digitality is a cultural condition and is not limited to the smart home space. At its broadest, living in digitality expands beyond the domestic sphere, to wherever people exist and interact with one another and ubiquitous, distributed, and connected technology. The smart home exemplifies living in digitality both spatially and symbolically, offering an ideal case study to examine the cultural condition of digitality. In the following sections, I briefly describe each key feature of living in digitality. I note its functionality and how it manifests spatially, and give particular attention to how the feature is articulated in the smart home space.[16]

Although these seven features are presented below as distinct manifestations of living in digitality, each is also imbricated within—and coconstituted by—each of the other functionalities. As such, they should be read as interlocking elements that constitute the human (read: social, political, and economic) and technical (machinic, code-centric, logistical) experience of living in digitality. For instance, in the smart home and beyond it, technological *optimization* and *computation* are each a side of the same digital coin: it is difficult to imagine optimizing the home without the computational processes that promise to enhance both domestic space and the home's relationship to the outside world. It is similarly impossible to understand the way that digitality seeks to *control* and capture human life without also understanding the way *bias* is built into the technodomestic machine. So, while these features are described individually, I encourage readers to imagine them as necessarily entangled and partial.

Table 1. Key Features of Living in Digitality

Key Feature	Function	Manifests spatially . . .	Smart homes (SH) rendered . . .
Optimization	Proposes and normalizes optimization of everyday life through technological processes	Through "efficient" arrangement of physical space; incorporation or overlay of digital devices that shift spatial capacity or function	As the next frontier for technological enhancement; subject to improvement through algorithmic processing; as a perpetual project of domestic achievement
Scale	Navigates multiple levels of scale	Through embedded or presumed interconnection of digital or becoming-digital environments	As nodal point between microsystems (individuals, families) and macrosystems
Computation	Constitutes lived reality through computers and computational processes	By classifying space according to its commensurability with devices, networks, sensors,	As worthy of individual and collective investment and value insofar as they are coterminous with symbolic value of computation
Generalizability	Transfers and expands scope of computation as the ideal for living, is structural in scope and generalizable beyond discrete technologies	By qualifying space, objects in space, and bodies in space as machinic, protocological, intelligible, and optimizable	As an ideal and possible domestic achievement, even when "smart" technologies aren't present or eminently visible
Control	Directs action, organizes imagination, figures possibility	By locating agency locally (e.g., the device or homeowner) but operating at the level of technoculture; obscuring influence of the digital	As an extension of the state, an enactment of bourgeois ideology, and a platform for domestic management
Bias	Bakes in discriminatory effects, obscures them through neutrality discourse	By reinforcing difference and imbuing dominant spaces with authority and value across multiples axes of power	By systematically prioritizing privileged digital subjects and smart device users at the expense of others, affirming architectures of dominance

Optimization

By virtue of their existence, smart homes suggest that domestic life is—and ought to be—modifiable through a constant, adaptable, and commodified process of technological innovation. On the consumer-facing side, this innovation is explained as a benefit of the smart home: to craft the most optimized life for domestic inhabitants, ideally through the acquisition and continued use of smart products. Culturally, optimization signifies totalized efficiency achieved through rote, quotidian quantification, and categorization of human and nonhuman activity. Living in digitality therefore is a cultural condition that produces the lived realities of the social and economic world by making it more "legible, recordable, and knowable" through "processes of capture, definition, optimization, and filtering."[17] Because it sounds positive, optimization provides rhetorical cover for technological hegemons' quiet grafting of extractive data processes onto increasingly intimate spaces. This comes as no surprise: in Silicon Valley and beyond, optimization has become a north star driving the development of everything from discrete consumer goods to entire cities.[18] Optimization, then, is not limited to individual products or singular consumer experiences. Rather, it is a collective, expansive vision for how to enact computational goals at every level of human life. In its most hegemonic expressions, living in digitality assumes in advance the sociocultural import and utility of optimization as a core human value. Having done so, it works to normalize and neutralize this valuation and embed digital technologies into all parts of life.

Smart homes promise a streamlined domestic experience, optimized through technology, for the express benefit of their users. Optimization in the smart home space is a synonym for logistical efficiency acquired through ongoing engagement with digital technology. Optimization is enacted as an instrumentalist domestic experience brought about by the protocological codification of daily routine and made possible by the acquisition of smart products.[19] Smart homes reflect and amplify optimization principles by prioritizing efficient technological mediation in the home's aesthetic and spatial configurations. Here, smart home enthusiasts champion digitality for its ability to manage everyday life, optimize living conditions, and rationally organize space and processes. For example, rooms are arranged to decrease the number of steps needed to complete basic tasks like cooking; rooms "work hard" or "work smart" if they condense time and labor across contexts and purposes with smart products. Smart technology directs or facilitates the optimization, linking embedded features like sensors and cameras with productivity principles to figure space according to capitalist logics of production. Space not only supports this optimizing process; it becomes an object to be optimized. Digital enhancement is both a sociotechnical imperative and an illusory goal; "improving" living through technological process is, strategically, a perpetual project.

In the context of smart home spaces, optimization is often defined in the negative, as a lack that must be filled by a connected product. The home—already a complex collection of objects, spaces, people, and ideologies—seems to offer no shortage of opportunities for improvement. Smart products are described by their makers as solutions to life's little (or big) problems: a lack of time, lack of certainty, lack of security, lack of fulfillment, lack of excitement or engagement. The technological solutions offered in response to this lack are myriad and varied. For example, smart surveillance systems mobilize monitoring, automation, and connected gadgets to promote a feeling of control. Smart cooking systems promise everything from precision cooking for established home cooks to adventures through cuisine for novice chefs to miracle-cure health fixes for the ill or exhausted. Virtual assistants are pitched as an extra set of hands or even a second brain for the harried digital subject. What holds these technologies together is a generalized cultural investment in optimizing tools as part of a techno-solutionism, even when the smart home is a "solution in search of a problem."[20] In theory smart products optimize the running of the home. They could therefore be expected to relieve the demands of labor, freeing digital subjects up to pursue community building, leisure, art. In practice these devices themselves require management and may increase expectations of domestic labor or cleanliness.[21] Furthermore, any temporary ease for smart home users is bought with more-intense, less-secure, and highly managed labor of the digital domestic other, or those who must participate in this economy of optimization with few alternate pathways for survival.[22] None of these smart logics encourage cautious introspection about the cultural conditions that constitute these problems in the first place. Rather, optimization is such a powerful concept that it seems to undercut any criticism at the quick. After all, who *doesn't* want to make life better, easier, and more efficient?

Optimization isn't (only) a metaphor referencing technological solutionism: it is a heuristic for evaluating what parts of life are important and useful and what parts of life need to be reworked by the digital. And, from the perspective of digitality, it *all* needs to be reworked. Domestic spaces may be called smart if they are orderly and intentional. Those parts of domestic space that are *not* orderly and intentional ought to be transformed through a series of digital mechanisms and technological discipline. And yet smart technologies are not limited to the domestic space. As Yolande Strengers has written, "the technocratic, functional and efficient ideals of the smart tag extend to people who use data and technology to mediate and moderate their behavior . . . 'smart' is not only used to describe ICT-enabled things, cities and countries; it also constitutes a distinctive ontology in which smart technologies perform and establish a highly rational and rationalizing form of the social order."[23] In other words, smart homes constitute and transmit a way of being that proposes and then normalizes optimized human existence.[24]

The logistical appropriation and persistent quantification of life is a key component of optimization. From the perspective of logistics, "optimization" refers to the perpetual drive for both increased efficiency and ever-increasing access over the human and nonhuman domain—the self, the social, the body, the material world, and collectivity.[25] Although logistical thinking is touted as a new technique for achieving optimization, it is in fact a historical ontology that has ordered our frameworks of thought and value for centuries. Harney and Moten, for instance, note the genesis of logistical processes and thinking in the transatlantic slave trade.[26] Logistics and its cousin optimization are therefore pre-digital. However, digitality increases logistics' capacity to "move objects and move through objects."[27] Logistics' ambition is as infinite as it is ruthless; in its drive to orient and optimize reality, it demands quantification, qualification, connection, and circulation of both subjects and objects—in physical places, in figurative spaces, and in the collective psyche.

Optimization necessitates the organization of social and political relations through an instrumentalist, computational framework—one that existed even before we understood computation as such. Optimization's systemic social force is part of a larger drive to embed and normalize what some have called platform capitalism and others have called data colonialism.[28] Despite differences in name, these concepts gesture toward the interconnections among data, capital, systems, and power; they signify a social, political, and economic imperative to appropriate human life and ready it for capitalist consumption. Couldry and Mejias have described this phenomenon as "the capitalization of human life *without limit*."[29] As Couldry and Mejias put it, "our activities and even inner thoughts occur in contexts in which they automatically *are made ready for* appropriation as data."[30] The smart home is one such context.

The other side of optimization is often exploitation. Smart homes do not remove the need for domestic labor, management, or other logistical processes that make the home run efficiently. Rather, smart homes simply displace them onto other humans or nonhumans (e.g., bots or virtual assistants) that are less visible to (and frequently less fortunate than) the smart home consumer. Put simply: someone or something inevitably pays for the privilege of optimization. As we will see, smart homes are domestic spaces prepared in an algorithmically mediated service economy for privileged digital subjects. In one of the homes we will examine, for instance, part of the luxury accommodation in a new-build home is a room called a "drop zone," a secure but separate space for contingent workers to deliver groceries and other items without engaging with homeowners. For homeowners, then, optimization offers ease and efficiency—a privileged temporality that is experienced by individuals but made possible by others.[31] This example also demonstrates that smart homes connect individual residents with vast material, temporal, and social infrastructures. In other words, they operate at scale.

Scale

In the case of the smart home, living in digitality appears to be an individualized experience with technology, set in the bounds of a discrete private space. Yet smart homes exert influence across society. This is true even when an individual user appears to "opt out" of smart technology. Smart homes connect individuals with broader social, political, and economic systems, logistically and culturally. Logistically, smart homes use smart products to create an experience of living in digitality by managing homes' historical connection with the public and private sphere. For instance, a key function of several smart home products is to manage and/or optimize the use of public utilities like water, gas, and waste for smart homeowners and residents. Smart homes subsequently become a nodal point with vast infrastructures beyond the individual homeowner's control. Culturally, smart homes tap into existing social and economic structures and contribute to their reiteration in technological form. Algorithmic processes on external platforms mediate the relationship between individuals and, for instance, delivery drivers dropping off food or home goods. As a framework for evaluating the social and political effects of smart homes, the theory of living in digitality examines both these small, discrete actions, and the system-level changes that occur when domesticity is processed digitally.

Smart homes are important battlegrounds for platforms, vast physical and digital infrastructures that have a significant pull in the social, political, and economic spheres. Nick Srnicek has argued that platforms such as Amazon, Apple, and Google are "consolidating monopoly power . . . as they collect immense wealth."[32] Platforms are integral to contemporary iterations of capitalism, in part because they are flexible, dispersed, and robust. As Srnicek suggests, "rather than having to build a marketplace from the ground up, a platform provides the basic infrastructure to mediate between different groups. . . . Platforms are, as a result, far more than internet companies or tech companies, since they can operate anywhere, wherever digital interactions take place."[33] The smart home is a lucrative site for platform capitalists insofar as it imbricates their residents—and their daily habits, preferences, activities, desires, wants, and needs—within the scale and function of a platform. Tech journalist Dieter Bohn has suggested that smart home tech, and in particular smart virtual assistants, could foment an all-out "platform war" between major technology companies vying for space in the modern home.[34]

By creating an experience of always-on, ambient, technological living, living in digitality "plugs" the domestic sphere into the greater digital economy.[35]

In this digital economy, data—in all its forms—is capital. And that capital scales—up and down. Smart homes connect their residents to technological and economic infrastructures, subtly and carefully stitching users into the platform economy from the comfort of their homes. In exchange for novel domestic services, smart homes supply crucial information about users and their habits to the platforms that help smart technologies run. From this perspective, the home itself becomes a platform, and "domesticity . . . is formatted to run on the respective smart home operating system."[36] Smart beds can track users' sleep patterns, send that information to the cloud, algorithmically process it, and offer personalized sleep suggestions. Smart refrigerators outfit the fridge interior with cameras so that users can monitor their groceries. The fridge's Wi-Fi connectivity and third-party app integration mean users can order milk right from their fridge. Smart technologies certainly serve their users, but they also have other masters, including the major technology corporations that manufacture, market, and sell these smart devices. Paradoxically, then, smart homes uphold normative social, political, and economic structures even as they help cement new economic formations, including platform capitalism, a modern arrangement in which capitalist economies are organized not through manufacturing or other sectors but through the infrastructure of the platform. In this way smart systems are productive tools for technology corporations in a digital economy organized around data accumulation, processing, and transfer.

Digitality's social force operates at the level of system while rhetorically scaling that force toward the individual. The smart home purports to place control of one's physical surroundings in the hands of consumers. In reality living in digitality offers only a handful of levers for smart home users to pull. When living in digitality, the smart home is signified as an intimate space in which technology hegemons, government entities, and other powerful agents can exert influence. Smart homes are not inert domestic spaces or products. Rather, we should understand the smart home to be a complex assemblage of technologies, spaces, and discourses operating across scalar dimensions, drawn together under the shifting sign of digital advancement.[37] In other words, smart homes distribute the technical and cultural logics of computation, connecting the intimate interior of the home to the domestic exterior.

Computation

If living in digitality operates at scale, computation is the discursive and technical operating system that makes this scale possible and palatable. At its most basic, computation refers to the idea that we can and should strive to make life more like the computer: logical, ordered, mechanistic, technical, mathematical. The ubiquity of the computer (in our fridges, in our pockets, on the sidewalk) seems to validate the idea that computation is inherent to life in a modern society. And so, in one sense, computation refers to the fact that more and more people own, engage with, or interact with computers embedded in more and more objects. But, in another sense, computation gestures toward something more abstract and structured: that society is moving increasingly toward the machinic logics endemic to computational society that operate beyond any individual object. It is not enough to say that there are more computational devices in the world than there were ten, fifteen, or twenty-five years ago. Rather, when I refer to living in digitality as computational, I mean that computation increasingly structures both our engagement with the material world and our political imaginaries about what is true, real, available, and valuable. Computation, then, is not only a descriptive signifier but also a normative one.

Computation is a logic that both describes an ideal world and provides the means to draw that world nearer. Computation may refer to a machine, a tool, or a device that is electronic, automated, or otherwise technological. But computation is also a process and a framework for engaging with the human and nonhuman worlds. The computer, having been built by humans, is necessarily a cultural object, with cultural effects that orient humans to life beyond the individual object. Given the computational framework of living in digitality, spaces are valued based on whether or not they are coterminous with either individual computational devices or with the values of computation more generally. From this point of view, living in digitality should be understood as a cultural technology that mediates users' engagements with a reality computation helps to construct.

If living in digitality encourages individuals and collectives to manifest the optimal through digital logics, the computer is the framework through which those logics are operationalized. Here, the computer is a tangible object but also an overdetermined cultural symbol. Digital technology scholar Seb Franklin has theorized digitality as a cultural logic organized around the computer as a tool and a metaphor.[38] Computation proposes a set of seemingly unassailable values against which to measure and manage the world. Order, rationality, logic, and quantification provide a framework for measuring one's relationship to the social and political sphere. At the same time, the social and political sphere can be captured, calibrated, and reorganized into the vision of computation, which is always already assumed to be the ideal for living life with others. Digitality takes this orientation and intensifies it, mapping the protocological experience of computation onto the messier experience of humanity. In this way, computation serves not only as a vision for building society but also the particular tools, processes, and ways of thinking necessary to make this certain way of living possible. When living in digitality, computation moves beyond its presumably inert nature as a tool to manifest the technological reality underwriting digitality.

Living in digitality is organized under computation as sign. Ultimately, computation is a productive force; living in digitality is productive of a new domestic subjectivity in the form of an optimized smart home user that expects and even demands that all aspects of homelife be mediated by computation.[39] For the smart domestic subject, the smart home is the impetus for and a space in which to enact these principles in familiar, intimate spaces. Here, computation gestures toward the agential capacity of smart homes to prepare current and future subjects produced for a digitally mediated present and future.

A key function of smart technology is to record, process, organize, and mediate everyday life. When smart technologies are built into the home space, the smart home encourages residents to live according to computational values. Smart surveillance cameras, like Amazon's Ring or the Google Nest Camera, encourage homeowners to document, survey, and secure the bounds of the smart home. Smart appliances encourage increased domestic productivity by drawing on the dual signifiers of optimization and consumption. Smart beds track and claim to improve sleep (and sex and work). Smart refrigerators predict residents' nutritional needs and project algorithmically perfected shopping patterns. Smart locks enable strangers to enter the home to provide platform-enhanced on-demand services, and more. In all these instances, smart technologies help process all aspects of domestic life through computational logics as well as through individual devices.

Computation manifests spatially in the smart home in multiple ways. Most generally, computation helps qualify spaces as meaningful and useful insofar as they are commensurate with digitality as a logic. In this way, computation qualifies through quantification; quantification, in turn, draws on a legacy of spatial optimization and logistics as a way of figuring the value of the invaluable (or, perhaps more concretely, the incalculable). Homes become laboratories (sometimes literally) for new technological imaginaries. Domestic spaces are made ready for increased connection, datafication, and quantification. The digital devices that make up the physical and material constellation of computation at home not only mediate existing spaces—they produce them. The introduction of smart technology brings home computational logics by encouraging a set of actions, mindsets, and frameworks that align with digitality as the domestic ideal. For instance, the incorporation of a smart speaker enables users to literally and metaphorically rewire existing spaces for new purposes. Smart technologies can also prompt novel relationships with space mediated by Amazon, Google, or Apple. Smart sensors turn the home into a site of perpetual and owner-operated surveillance. This data then may be used by parent companies to construct new, rich maps of the human experience at home.

Smart appliances rework humans' engaging with age-old practices of cooking, cleaning, and care. In so doing they tether and untether the domestic space to and from external infrastructures, including supply chains, labor markets, public utilities and more.

Like other domestic spaces, smart homes communicate the social values of societies in which they are imagined, occupied, and used.[40] Since "buildings can act as a repository of our ideals," smart homes mark and (re)entrench cultural values.[41] Like other homes, the cultural value of smart homes transcends the glass, wood, brick, concrete, drywall, adobe, wiring, and other materials from which they are constructed.[42] Spatially, smart homes are the manifest expression of a pro-technology cultural imaginary. Culturally, smart homes modify traditional concepts of domesticity by intensifying the computational and social logics that organize our social, political, and economic realities.

Smart homes are evidence that modern domestic spaces are being built in alignment with dominant narratives about technological progress. That is, smart homes represent in physical form discourses about the imperative to live with technology on a daily basis. Consequently sometimes the meanings of domestic spaces change with the incorporation of smart technologies. However, equally as often the physical layout of spaces can adapt to work best with networked technologies. As these shrines to computation are built or modified onto existence, smart homes help to (re)configure life at home to be more seamless with technological innovations that are happening outside of them. Said another way, smart homes enact technological values and discourses within living spaces themselves.

Generalizability

Smart homes are generalized computational environments in which technological products and processes universalize the principles of living in digitality as a matter of daily practice. Living in digitality, then, refers to a collection of sociotechnical tools, practices, spaces, and imaginaries that encourage, normalize, and justify the sociotechnical values of optimization, information processing, and technological configuration *in all parts of life*. While digitality is fundamentally about "digital information and its processing and transmission," it is also about the social, economic, and political conditions that organize life according to these technological logics.[43] Smart homes intensify the logic of computation by enrolling consumers into not only a "set of technologies or logical operations but also a fundamental condition" of digitality.[44] Computation becomes the basis for configuring digitality, and its justification for doing so in a robust and ongoing manner.

On one level, then, living in digitality can refer to everyday engagement with individual devices, including electronics, computers, smartphones, and other gadgets that constitute the smart home. Yet living in digitality is not limited to a set of tools or technologies. Digitality is a *cultural logic* that organizes and orients engagement with the world.[45] As a distributed and pervasive social force, digitality is "a set of technical principles . . . that persists beyond any specific device or set of practices."[46] Digitality is generalizable.

Unbinding technology from individual devices clarifies how the logic of the digital manifests across human experience, including domestic spaces. We can understand living in digitality as an *epistemic* condition coordinated through the expansion of computation, control, and technical logics along two axes. First, digitality presumes to make modern (home) life possible through the systems, logics, and technical configurations of computation. Here, digitality functions as a well-distributed set of processes or tools; in this way, digitality is equipment for living according to computational principles. Second, living in digitality puts these tools to work under a broader sociotechnical condition wherein the computer is figured as both an object to engage with and a metaphor for how society might optimally function.[47] As a metaphor, computation figures the social as augmentable by technology and by technological thought. More than that, generalized computation, as a desirable framework for living, structures modern political imaginaries, in part because it reenvisions its constitutive parts as generally technical. Bodies and beings are understood systematically from a computation and information-processing standpoint; systems-level phenomena like governance and economics are presumed to function mechanistically. Value is afforded along protocological lines. Systems and objects that do not fit within this ideological structure can be brought online with the right tools and thinking.

Smart homes are the manifestation of what Franklin has called the "becoming-general" of digitality—harbingers of a generalizable experience of technocultural control and capture at the level of environment. The smart home affords residents an opportunity to engage with technology in an embedded and expansive way. Rather than interacting with *discrete technologies* (e.g., devices, sensors, appliances, and so on), the promise of the smart home is that digitality will be made available as an *environment* in which to live, work, and interact. Smart homes make an argument for digitality as a framework for organizing everyday life. As a generalized expression of digitality, smart homes invite their users to exist within a spatialized version of the digital imaginary. In so doing, smart homes help to reconceptualize the relationship between space and technology.

The smart home distributes technological artifacts and processes across physical space, enrolling its users in an ongoing, totalizing process of technological placemaking. Digitality domesticates space even as spaces help direct the possibility of digital mediation in real time. The smart home translates domestic space into a novel form of technological place by articulating the role of technology in everyday life. In its wake the smart home modifies, reinforces, and challenges social structures around the domestic space and the experiences of the people living within it. Living in digitality spatially manifests the "progressive opening up of human life to externally driven data extraction . . . the capitalization of human life *without limit*."[48] From this perspective, smart homes are only the latest articulation of a technological condition that colonizes space (and the people within it) in the name of abstraction, extraction, and quantification.[49] Understood from this perspective, living in digitality represents the spatial enactment of digitality but also its repeated folding into domestic spaces in material and discursive ways.

The becoming general of technology is an important component of living in digitality because it prepares the messy intangibility of everyday life for data extraction. As Couldry and Mejias note, "human life first needs to be configured so as to 'naturally' generate a resource such as data. Momentary data about one individual's actions or properties is worthless unless it can be combined with data about *other* individuals, actions, moments, and properties to generate relations between data points."[50] What is not yet ready for computation can be made ready through various sociocultural and technological processes, because value can be derived from generalizable settings or experiences only through these cascading processes readying life for capture. Smart devices help to capture the generalized horizon of human activity. Having captured it, these devices may also render the experience quantifiable in the form of data. The smart home generalizes digital technology as a metaphorical and literal landscape in which one might exist. In so doing smart homes recalibrate user engagement with technology toward productive (and extractive) data practices.

Control

Smart homes are novel and useful insofar as they help users modify and manipulate the spaces around them. The primary user-facing functionality of the smart home is to equip users to control the domestic environment. This form of mediation purports to put technological power in the hands of users. And smart home users sometimes do have the power to control certain aspects of homelife. As noted at the opening of this text, smart technology can empower users to cook food in a certain way, monitor spaces and movement within those spaces, and communicate with other humans through wireless technology. This can be understood as a form of individual control made possible by harnessing the affordances of the digital to complete discrete domestic tasks.

But just as often, smart technologies create only a *sense* of control for smart home users. In reality, the terms of agency are constrained by processes and procedures of living in digitality. When depicting living in digitality, it is useful to be precise with our definition of control, as it implicates what we might mean by agency (and the ability to resist or counter the controlling regimes of digitality). For our purposes, control is a complex social force that refers not just to individual action or the autonomy of self but also to a range of social practices of varying intensities that enable or disable movement, creation, imagination, and collectivity. Franklin reminds us that digitality is characterized by (and circuited toward) control, organization, optimization, and extraction. Living in digitality is only the most recent expression of the "control episteme," which was "established as a central principle upon which ideals of informatics, always-on, and globally distributed modes of organization and valorization are premised."[51] As a set of technically ordered knowledges and practices, living in digitality mechanizes control by organizing human relations under the sign of computation, which is rendered both principle and process of managing life.

Here, *control* alters the social fabric by orienting life toward the machinic, the optimizable, the computational, and therefore the hierarchical. Control modifies what Couldry and Mejias call "the minimal integrity of the self," a fundamental essence of humanity that is currently under threat from data colonialism.[52] Data colonialism is a set of social and technological relations premised on the violent appropriation of human life.[53] Data colonialism is a social force oriented around control, one in which "ever more of our activities and even inner thoughts occur in contexts in which they automatically *are made ready*."[54]

Control manifests protocologically on individual bodies through the techniques of digitality, but individual actions can support or resist these protocols. With the aid of smart technologies, bodies may be monitored for performance and efficiency, replicated as data doubles, (re)produced as productive/consumptive domestic subjects, and/or positioned as rational beings encouraged to barter their individual or collective capacities for apparent control over their environments. Here, smart technology enables and disables according to one's status as a digital subject. When living in digitality, those digital subjects with the least privilege may have the least amount of control over their bodily autonomy. For instance, workers in low-wage, supposedly low-skill positions (like "pickers" and drivers for Amazon) are more likely to have their bodies monitored on a second-by-second basis to ensure efficiency, and therefore profit.[55] Others on the margins may be computationally marked as a threat and subsequently surveilled and (sometimes violently) securitized. Others with the most privilege may be granted some level of control over the terms of engagement with technologies. They may also be able to determine the status of others with respect to surveillance, extraction, or datafication.[56]

Control is endemic to living in digitality but not limited to the creation or use of digital technologies in domestic space. Nor is control vulgar or unidimensional in its effect on the individual and the social. Rather, when living in digitality, control manifests in banal, often unnoticeable ways—particularly in spaces we frequent every day. When a resident uses a smart appliance on a daily basis, they may or may not consider the intricacies of development and connection necessary to make the device useful. A smart fridge that connects to a grocery-ordering platform, for instance, not only mediates the individual users' experience with the refrigerator as device but implicates systems-level processes in the production, acquisition, transportation, and delivery of food. The idea of computation as generalizable beyond discrete technologies is vital for understanding the scope of digital control. In the context of digitality, Seb Franklin defines control as a "logic under which social worlds are reconceptualized as information processing systems."[57] Franklin's definition is useful insofar as it articulates the structural functionality of both digitality and control. Here, digitality and control are systemic and thus well distributed. They refer to individual engagements with technology insofar as they exist within a spectrum of prescribed and normalized technological practices at the level of the social. But beyond that, control is pernicious because it creates the conditions of possibility for knowing about and engaging with the self and the social. Sometimes control manifests individually; here, control can refer to "material self-regulation" along with the tools, techniques, and infrastructures that make self-regulation possible.[58] And yet control is a much more widely distributed process that is ideological at the level of structure. Living in digitality, then, implicates the individual but clarifies that any choice offered by digitality is constrained by social and technological conditions of the social.

Smart homes help domesticate digitality as a systematic force, concretizing in daily practice the "very digital logic that undergirds both the technical and sociocultural organization facilitated by control, from computer hardware and software to the conceptualization of thought life, and social interactions as inherently digital and valorizable."[59] These technological logics are motivated by technology with agency, "creating systems that monitor, manage, and manipulate" smart home users, smart technology consumers, and even those with no discernable relationship to the smart home.[60] Smart homes direct people and society on how to interact with the home and with one another. Moreover, smart homes invite powerful actors from the social, political, and economic sectors into the dining room, the bedroom, and the bathroom, where they can receive unprecedented access to the intimacies of humanity. In sum, smart homes invite external influence and control into the domestic space through algorithmic protocols, technological affordances, and socially constructed technologies. Put simply, living in digitality is the domestic experience of technological control in the home space.

So, too, the domestic sphere creates, upholds, and intensifies cultural norms, patterns, and expectations of the home as an institution and organizes how the people inside of it should look, act, and be. For instance, scholars have shown how domestic architectures uphold stereotypical gender norms, intensify heteropatriarchal expressions of family life, and assimilate the middle and working classes into an oppressive version of civil society.[61] Smart homes continue this tradition of cultural creation but escalate its effects through what is often called technological innovation. In truth, both innovation and the smart home are limited by the social and economic intensities of digitality, namely, the focus on instrumentalizing, optimizing, and digitizing all aspects of human life.

Smart homes are spatial emblems for digitality, and they manifest communicatively through discourses of control. To be clear: any control over the technological or physical space by otherwise privileged digital subjects is partial and therefore incomplete. But the smart home plays otherwise: control and capacity are available for purchase if only one would buy into the smart home ideology. Like all ideologies, this one presents with contradiction as pertains to control. Many smart home users sense that these technologies are not totally under their control. This is why, for instance, pro-tech media in the smart home space must do double duty in crafting the digital imagination. At once, smart home media must argue for the utility and agency of smart devices to make a meaningful difference in the home while also convincing users that they will not lose their sense of control over the domestic space, which socially has been regarded as a place of refuge and security.

When discussing the smart home, control is a complex subject. Although individual smart homes serve individual residents, they are collective cultural achievements with broad cultural effects. Smart homes are particularly fruitful for launching projections of a digital future, because they tether the cultural capacities of domesticity with the agential nature of computation as a framework for living. To fully understand the force and effect of control as a key component of digitality, we must first consider how digitality is built, used, and normalized within hierarchical systems of power and value. We must examine how living in digitality is biased.

Bias

Individual experiences of living in digitality are not universal in effect or by design, because technology is not universal either in effect or by design. Rather, smart technologies are organized by systems, structures, and logics calibrated differently depending on one's identity and social location. These calibrations activate and extend long-standing and newly developing sociotechnical hierarchies. When living in digitality, one is likely to experience technological bias and control in varying forms, intensities, and expressions depending on one's status as a digital subject.

Scholars of race and technology offer useful analytical tools for examining bias as a function of living in digitality. In *Algorithms of Oppression*, for instance, Safiya Umoja Noble challenges the "benign instrumentality of technologies," analyzing how everyday technologies like search engines are organized around—and actively contribute to—hierarchies of oppression.[62] Black digital studies scholar André Brock argues convincingly for "beginning science and technology studies from racial and cultural grounds rather than limiting analyses to the technologies themselves."[63] Choosing to examine novel technological developments such as the smart home through a cultural lens attentive to power and difference challenges hegemonic intellectual assumptions about smart technology across several axes. First, as Brock notes, such an approach is a fundamental critique of technological instrumentalism or the idea that technology is an inert set of tools that is "value-free . . . neutral, meaning that it has no preference as between the various possible uses to which it can be put."[64] Second, as Brock notes, a racial-cultural approach to science and technology rejects the assumption that the development and use of technology are color-blind.[65] Rather, because "race has always already been an informational group identity, designating class and cultural capital (or the deliberate denial thereof)," living in digitality is by definition always already raced, gendered, and classed. Human tools are shaped by their makers' politics and biased principles in ways both anticipated and unanticipated.

This inherently skeptical orientation to smart technology is particularly warranted because makers and users often represent smart homes as value-neutral spaces composed of disinterested tools that can be used for society's benefit. For instance, in chapter 1 a smart home ad refers to the future as "how we're going to live. All of us." This perspective tells only part of the smart home story. What is left out are the ways that smart technologies—like all technologies—are shaped by what bell hooks has called "white supremacist capitalist patriarchy" or the "interlocking systems of domination that define our reality."[66] Technology is not "removed from political concerns and consequences. . . . Instead [we must] recognize the politics that have been, and continue to be, an integral part of technology all along."[67] Smart technologies are not the only non-neutral cultural force at work in the smart home. Rather, as I show here, smart homes are predicated on extending a domestic ideal that, historically, has been accessible only to individuals with capital and social, economic, and political privilege. For instance, in the American context, domesticity has historically been categorized by dominant classes through technological grammars of power, difference, and alienation.

The history of American domesticity is revealing on this point. Modern conceptions of domesticity in the US are often attributed to a cultural break around the time of the industrial revolution, which featured major technological advancements, changing social configurations around publicity and privacy, and seismic shifts in economics and labor. These changes were filtered through concepts of home, forming a new domestic foundation for nineteenth-century values. The domestic ideal was (and continues to be) organized around the mythic figure of the ideal woman—presumed to be white, middle- or upper-class, a mother, and Christian.[68] The mythos of dominant domestic femininity functioned as a productive technology all its own. A key function of this ideal domestic subject was to perpetuate a normative ideal through maintenance of the house, instruction of the family, and protection of the normative domestic worldview. In this domestic ideal, home is considered separate from the messy world of the public sphere—an assumption that persists today, albeit in different ways. In the mythic domestic imaginary, home—and those who are in it—are frequently positioned as epistemological and ontological adversaries to those who exist beyond the home.[69]

Historically, domesticity is a technology that offers a discursive reservoir for anti-black violence, linking the original violence of slavery to "stand-your-ground" and "castle" laws. Modern American conceptions of smart homes as gracious, efficient, and productive owe a great deal to a racialized nostalgia for a not-so-distant era in which the household was run using the unpaid labor of enslaved people or else serviced by low-wage workers who did not experience domestic life as distinct from the public world (of work). Domesticity as a technology for racial oppression in the United States provides both mechanisms for violence and logics that justify its use. For instance, Hortense Spillers has argued that "the horror of slavery was its absolute domesticity that configured the 'peculiar institution' into the architectonics of the southern household. So complete was its articulation with domestic economy that it loses visibility and becomes . . . 'natural.'"[70] Indeed, the logics of slavery predate and articulate modern domestic efficiencies later used to justify the smart home. Couldry and Mejias note that "the treatment of human beings as mere property stimulated the rationalities of profit maximalizations, accounting precision, and data optimization that we now tend to associate with modern rationality."[71] In her treatment of the "uncomfortable mathematics of black life" and the "violent arithmetic" of the transatlantic slave trade, McKittrick makes a parallel claim, noting how rationalities of computation are epidermalized—literally inscribed on the Black body in the form of lashings on and en route to the plantation.[72] The inhumane rationality of the plantation demonstrates how modern domesticity can use grammars of both rationality and gentility to obscure the violence behind the domestic ideal. To this day, companies hawking that mythic domesticity across a bevy of products use terms like "*plantation* in their names to evoke a romantic nostalgia for a bygone era of sprawling mansions, women in extravagant corseted gowns, and Southern gentility, while ignoring the brutality that accompanied it all."[73]

The smart home's focus on efficiency and optimization can similarly devalue domestic labor historically done by poor women and by women of color. As Thao Phan notes, smart domesticity perpetuates a nostalgic form of "idealized domesticity" that is "reminiscent of the ways in which early twentieth-century middle-class households used domestic servants to resolve certain contradictions between feminine virtue and the hard labor of domestic work."[74] This undercurrent of domesticity is persistent and resurfaces in the domestic vernacular of the present day. Domesticity holds steadfast to its conditioning and projects its path into a smart home future—it often calls on complex, routinely conflicting discourses of ease and leisure inside the home and threats outside of the home.

Later, violent racial codes were rearticulated into domestic life through de facto and de jure policies, processes, and procedures that limited people of color's access to both housing and the social safety of domestic life. Redlining, for instance, is a set of legal, economic, and cultural practices used to segregate both poor folks and people of color—or to maintain segregation in communities where it already existed.[75] An early iteration of domestic digitality, redlining deployed supposedly objective algorithmic thinking to justify the disinvestment of prospective homeowners who are poor/working-class or nonwhite. The oppressive redlining system was self-referential in its logic and discourse. Rhetorics of (credit)worthiness, paired with bias baked into the sociotechnical structure of lending and financing, made both overt and covert discrimination appear both objective and logical, although it was neither. Although redlining was outlawed by fair housing laws in the middle of the twentieth century, ongoing practices of racialized disinvestment in communities continue in other ways, justified by the same algorithmic logics that undergird the smart home narrative.

The same smart technologies can be experienced differently by different bodies, in part because they were not imagined or built by all of us. Living in digitality is a cultural experience that may alienate us from the shared world of the public by locating us in very particular configurations of digitality.[76] These configurations vary according to a number of factors, including our race, gender, self-presentation, class, ability, access to digital infrastructure, and more. Access to sophisticated technology is disproportionately distributed to powerful global economies and those with the most power within them. Studies have shown that marginalized or minoritized communities may have access to digital technologies, but they may be lower quality, more expensive, more difficult to access.[77] This exclusionary structural phenomenon is called "digital redlining," in which "communities' differential access to digital technologies that is created in coordination with internet service providers and technology companies emulates banks' and lenders' racialized decisions about extending capital and economic investment."[78] Digital redlining echoes historical redlining in its logics of procedurality, which mirror the sociotechnical infrastructure of the smart home.

Marginalized or minoritized communities are imbricated in a vast network of digital surveillance, capture, and control that disproportionately affects their lives and livelihoods. For these populations, rather than being "technology poor," Virginia Eubanks notes, "data-based systems were ubiquitous in their lives, especially in the low-wage workplace, the criminal justice system, and the public assistance system . . . high tech economic development was increasing economic inequality . . . intensive electronic surveillance was being integrated into public housing and benefit programs, and policy-makers were actively ignoring the needs and insights of poor and working people."[79] Disparities in access, agency, and choice are particularly pronounced in the housing space, where "security regimes" are well distributed across populations but with vastly different effects.[80] Surveillance studies scholar Torin Monahan has argued that while affluent gated communities share "remarkable similarities" with lower-income public housing in terms of surveillance, the direction of the surveillance is distinct.[81] Whereas gated communities can leverage technology to surveil possible threats, in public housing residents are the perceived threats. For Black and brown communities, as well as other communities deemed suspect by society, living in digitality often means being enrolled forcefully into a surveillance society with technology controlled by others.

These discriminatory patterns persist today in the smart home. When smart home residents opt in to using devices like Amazon Ring under the guise of individual, familial domestic security and "peace of mind," they are implicated in a broader skepticism of the outsider. Katie Lambright has argued that "the restrictive housing ordinances of the midcentury contribute to the racial segregation still reverberating in the present. . . . The housing discrimination also fed into fear and distrust of the 'outsider.'" Ring is a technological response to the constructed social problem of the "outsider."[82] It relies on a vast physical and ideological infrastructure that conjures up fear of the other to sell products for the "safety" of a community, with or without the explicit permission of individuals within that community whose image may be captured by the Ring device. This infrastructure, which includes the Amazon platform and police partnerships, not only implicates others in the smart home user's proximity—it can put them in grave danger because of racial stereotyping, predatory policing, and the historically racist hypervigilance of vigilante neighborhood watches. As a surveillance device, Ring gestures to a structural, historical problem rooted in housing segregation and discrimination in America.

As discussed earlier in this chapter, smart homes also contribute to economic bias. Presently, users' experience of living in digitality is organized through platform logics imbricated within larger economic infrastructures.[83] These infrastructures, in turn, recreate socioeconomic divides between smart homeowners and those on the margins. Even experimental smart homes that endeavor to undo unjust social, political, and economic infrastructures are constrained by manufactured conditions of austerity, social unrest, an eroding safety net, and political gridlock. In this way smart homes can provide the good life for some, but not all. Elevated domestic experiences of the good life are often made possible by people on the margins of society who labor, to benefit privileged classes. For these less privileged individuals, technology extracts profit for platforms through efficient monitoring and surveillance.[84] At the same time, these surveillant capacities are domesticated through the gendered and raced personas of artificially intelligent virtual assistants such as Siri and Alexa.[85] Living in digitality is a cultural condition that is not extended universally, nor is it paid for equally.

Although smart homes are frequently depicted as futuristic, they still traffic in regressive narratives about what home is, who should live within it, and what their roles are. As the case studies in this book show, many smart homes reinforce larger cultural discourses about home as a place of social (re)production, where individual families live and work in many of the same ways as earlier generations. What is different about the modern smart home is the intensity of smart technology as a mediating force for these traditional cultural scripts. Whereas emerging technologies could materially and spatially reorganize life in the modern American household, the smart home brand prefers instead to maintain domestic traditions by routing them through the technological infrastructures. Practically, we can understand smart homes to be meaningful extensions of older ways of living with slight technological adjustments for consumers. However, behind the scenes, the smart home is an important medium for consolidating power, manufacturing demand, encouraging consumption, monitoring populations, and encouraging certain domestic behaviors.

The smart home is a tool for social control. As such, smart homes cannot be conceived as inert technological artifacts—in part because technologies themselves are social constructions that bake in power dynamics from the moment they are conceived to the point at which users take them up. Here, digitality is understood as a biased cultural force with persuasive capacities—it can influence how humans engage with the world based on dominant ideologies. At the same time, digitality is a product of cultural invention—the ideas, concepts, and tools associated with digitality come with cultural baggage built in. Smart homes are pitched to a particular consumer, "cast in the male-dominated industries of engineering, economics and computer science—and imagined in highly functional and masculine ways."[86] At the same time, the smart home turns on gendered stereotypes through "feminized artificial intelligence, digital assistants and robots that take on traditional wifely roles in the home . . . to provide a more acceptable and seemingly 'caring' form of surveillance in the home."[87] In addition to being gendered, the smart home is also profoundly raced and classed. Virtual assistants and other automated products are organized through "aesthetics of whiteness" that "romanticize the relations of servitude in a way that denies the pain and historical context of that relation."[88] In the smart home, everyday practices of living and the computational logics of digitality are non-neutral and configured by social, cultural, and economic forces that impart meaning, value, utility, and resources to groups in power. At the same time, digitality offers the discursive and material resources to keep those groups in power.

Incorporation

Living in digitality is the total, everyday expression of digital culture in the domestic sphere. When living in digitality, humans, devices, systems, and spaces are all incorporated by the social and technological systems of digitality. Incorporation is the process through which technologies stitch together the domestic interior and exterior, making it difficult or impossible to exist beyond the reach of digitality. Incorporation is similar to generalizability in its functionality but is distinguished by its attention to digitality's significant worldmaking capacity at the level of environment, subject formation, and human and nonhuman action. Whereas generalizability conceptually refers to how computational logics are distributed across spaces, incorporation refers to the ways that domestic spaces and the objects and subjects within it are constituted within the assimilating social and technological structures of digitality.

Smart homes encourage technological encapsulation, where users and others are habitually enmeshed within smart technologies across domestic spaces. Digitality is expansive, agential, and transformational. Scholars have noted that "smart technologies . . . have agentive capacities and the ability to transduce space—increasing the flows of material, labour, and capital through the home."[89] In exchange for increased control over the domestic sphere, smart technologies (and those who service them) promise what Sarah J. Darby calls a "domestic machine-utopia."[90] The flexibility and durability of domesticity as a cultural fixture carries always-on technological innovations into ever new terrains, from the home office to the kitchen, to the closet, to the bathroom, to a curious new part of the home, the aforementioned "drop zone." Scholars note that the home is "more than just a space where domestic and care work takes place. It is also an ever-changing assemblage of technologies. . . . The home is, in short, best understood as a means for social reproduction."[91] Living in digitality implicates domesticity, figuring the formation of subjectivity primarily by delineating the interior and exterior of the domestic ideal—what is included and what is necessarily excluded in the home space and outside of it.

As darlings of the twenty-first-century technological moment, smart homes have cultural influence that reaches beyond their individual doorsteps. Indeed, smart homes help shape the social, political, and economic realities of not only their occupants but anyone who exists within the cosmic pull of the "smartness" discourse. From this perspective, rapidly proliferating technologies are an inherently positive, inevitable occurrence. By its very nature, the smart home envisions a world in which technology mediates our lives 24/7/365 by encapsulating its users in an ever-complicated web of smart appliances, networked services, and technological infrastructures. Whereas, in the past, users were prompted to opt in to individual devices, services, and platforms (or some combination of these), smart homes create conditions where users are totally enmeshed in ambient technologies at all points in time. Along with their appliances and devices, humans' lived experiences may be "colonized by computer code."[92] What is more, the smart home goads the rapid expansion of digitality spatially by reorganizing the footprint of the home and shifting its contents.

When I first conceived this project and the concept of "living in digitality," I wanted a way to define the ways in which the digital seemed to creep around and inside the crevices of human existence. I wanted to explain the ways in which the digital—as a technological system and as a cultural heuristic—helped to architect a metaphorical and literal structure in which we all would live. In the smart home, the digital appears not only to attach itself upon the physical architecture of the home but constitutes its insides. To be sure, the digital functions as a tool that helps the home become smart and helps the residents of the home become attuned to a life lived with technology. In this sense the smart home enacts an experience of living *with* digitality. The preposition "with" seems to indicate a dual agency between humans and technology—a partnership that places each partner on equal footing. But this is simply not the case. Smart homes are part of larger sociotechnological infrastructures that don't exist as inert tools to be used by humans. We do not have total or always even partial control over the smart technologies in our homes. Smart technologies are crafted to appear as extensions of humans, to be wielded at their leisure, in their ultimate control. Instead, protocological digitality is all-encompassing, annexing more and more parts of the human experience for the benefit of data capitalism. Living *with* digitality isn't strong enough to describe how fully and totally integrated digitality is with human life. Living *in* digitality more forcefully describes the experience of capture we might experience within and beyond the smart home. Through the smart home, digitality becomes a place to live, a way to live, and an experience *in which* we are perpetually living.

Some of the influence wrought by living in digitality appear more or less benign for smart home users. For instance, some smart devices (like Wi-Fi-enabled appliances) gently compel their users to complete household tasks by engaging with particular platforms or networked services. There may be little difference between using Instacart, Shipt, or another grocery delivery service for the end user. Nevertheless, behind the scene, these actions correspond to seismic changes in the supply chain, logistics, and labor relations. All the while companies have access to users' shopping behavior, which "allows the company to serve targeted recommendations, predict purchases, and optimize prices on items (sometimes changing them multiple times a day), all to get people to spend more money."[93] For this reason, Jathan Sadowski has argued that smart homes are "data factories" in service of an ever-expanding, always-incorporating form of digital capitalism.

In sum, incorporation refers to the expansive processes through which much of human existence is mediated by smart technologies. In the case of the smart home, living in digitality manifests through the strategic and pervasive incorporation of a network of devices and services that encapsulate the user and the home, creating a curated media experience connecting the domestic interior to the outside world. When living in digitality, both space *and* imagination are subject to perpetual incorporation by technical devices *and* logics. In addition to architecting space through the logics of digitality, smart homes incorporate both users and nonusers by generalizing computation to life beyond the home. As shown in the rest of this book, the smart home's reach is so expansive, its generalizability so central to life in the modern age, that it is difficult to live outside digitality. Importantly, the decisions made by users who opt in to smart technologies have networked effects, constraining choice about opting out. Workers must service the home and smart home users (or else maintain the logistical lines that make smart services possible.) Networked systems of surveillance incorporate even anonymous passersby of the smart home into digitality. It is not that we are living *with* digital objects or their attendant ways of imagining the world. It is that digitality has manifested materially and spatially in such a total way that its reach is near complete.

SMART HOMES: TECHNOLOGIES TO LIVE IN AND LOGICS TO LIVE BY

Living in digitality is characterized by symbolic and spatial investment in digital logics of optimization, scale, computation, generalizability, control, bias, and incorporation. Although the features of smart home technologies may differ, what unites them is twofold: the persistence of embedded, ambient technology in domestic spaces and the cultural logics of digitality that orient life toward computational principles. Smart homes represent the latest expansion of digital technologies' reach into every part of people's lives.[94] Said another way, the smart home offers a fully mediated technological experience—a set of technologies one can *live in* and a set of sociocultural norms one should *live by*.

Processes of living in digitality are given meaning and direction by the signifier of the smart home, which projects both authority and cleverness—a way to manage domestic life and to exert agency on and through the technologies embedded in domestic spaces. Living in digitality figures expectations as well as the spaces themselves, such that the smart home intensifies the digital ideal, or the idea that every moment of our day should be mediated by digital technologies. In addition to presenting things as they are and have been, smart homes present the world as it might ideally be. In so doing, smart homes architect a version of lived reality in which digitality—in all its manifestations and effects—always augments the human condition. As chapters 1 and 3 evidence, smart homes stitch together traditional concepts of domesticity with slick, modern iterations made possible by explosions of platformed technologies. Even smart homes that challenge bourgeois conceptions of domesticity and familial relations (explored and explained in chapter 2) reify technology's central position in everyday life. Technological instrumentality is intrinsic to modern conceptions of digitality.

DEFINING SMARTNESS: A SOCIOTECHNICAL PERSPECTIVE

What makes a home *smart* depends largely on one's relationship to the home. For their residents, smart homes promise ease, comfort, and convenience. For the companies that make, sell, and service smart homes, they are strategic investments in a platform economy that deals in data.[95] For privacy and technology watchdogs, smart homes represent a rapidly encroaching form of digitality that makes everyday activities subject to algorithmic and logistical processing—and possibly security breaches. For gig workers who service the smart home through on-demand platforms and smart apps, the smart home is both an entry point into a digital economy and a site of potential alienation from it. For people of color, smart surveillance systems and related neighborhood watch apps can enable community care or encourage racial profiling. Countless orientations to the smart home remain, figured by conditions of living in digitality that imagine and produce digital subjects. Smart homes communicate conflicting values and expectations depending on audience and occasion. They do so spatially—by augmenting the space of the home—and discursively—through language, meaning, and discourses about domesticity and the future.

Those who make, market, and develop smart home technologies or build smart homes make a number of similar claims about the utility and value of the connected home. For instance, Brookfield Residential, one of the largest residential developers in the United States, has developed a set of smart home products designed around the technological empowerment of the homeowner or end user. "Forward-thinking innovation is here," the company notes, "putting complete control of the new home experience in your hands." Amazon, a leader in the smart home space, has deep investments in creating new use cases for their existing technologies; David Limp, senior vice president of Amazon, argues that his company "invents on behalf of customers."[96] However, much of this inventional action in the smart home space assumes in advance that smart homes are a solution to domestic life's problems. As Strengers notes, "the Smart Utopia proposes a world in which social disharmony and [other] problems are eradicated through new technology, without compromising current ways of life."[97] In truth, smart homes frequently produce a whole new set of problems in their wake. The glitz and glam of techno-utopian discourses about the smart home may make it difficult to take inventory of these problems.

At present major technology corporations have delimited what smart homes can be and do.[98] These hegemonic agents dominate the smart home space insofar as they control the smart device market. Amazon, Apple, Google, among others, are powerful and wealthy corporations with significant reach. Smart home technologies extend this reach into the domestic sphere, which is calibrated as a nodal point for the platform economy. Bringing smart technology home means that users are also bringing home the cultural, technological, and economic frameworks in which the smart technologies are imagined and built.

As such, smart housing scholar Sophia Maalsen has tasked scholars with considering the cultural and social phenomena that give rise to smart homes. In particular, Maalsen calls for smart home researchers to "better interrogate the range of stakeholders, ideologies, governance, market and material factors that shape smart homes."[99] An important step in this process is critiquing the implicit assumption that technologies are teleological or instrumental by revealing the hand(s) of their maker(s). It is not only that the companies pitching miraculous smart technologies have an outsized influence on the domestic space. Rather, the technologies they produce are pressed into service "as a tool to unproblematically bring rationality, efficiency and order" to the home.[100] Indeed, "technology is by no means a neutral bystander in the Smart Utopia. It is enrolled in performing a reality" in both the private and public spheres.[101] Investigating smart homes from a sociotechnical perspective highlights the agential capacity of the smart homes. This framework also explains how smart home technologies themselves are shaped by structural forces, including politics, economics, ideology, and more.

BOOK METHODOLOGY AND APPROACH

Threshold highlights the spatial and rhetorical force of smart homes and of the social phenomenon of living in digitality. The case studies, theoretical approaches, and methodologies used in this book operate at the intersection of rhetorical studies, digital media studies, infrastructure studies, studies of space and place, and critical theory. *Threshold* is a sociotechnical project that stitches together critical-cultural theoretical perspectives with a study of rhetorical practice in smart home space. Because homes spatially reflect larger public discourses about family, leisure, labor, citizenship, and more, this project leverages critical theoretical frameworks to unpack power dynamics at play in the smart home.

In the chapters that follow, I use primarily rhetorical tools to investigate how living in digitality manifests in smart homes, using optimization, scale, computation, generalizability, control, bias, and incorporation as primary entry points for my analysis. I assume that these facets of digitality are rhetorically negotiated at various levels of the social and public spheres, and that this ongoing negotiation occurs discursively but also spatially. As such, my methodological processes require strict attention not only to the materiality of rhetoric but also to its apparent ephemerality in physical spaces. First, I engaged in several years of rhetorical fieldwork at smart homes across the United States. Having situated myself within the physical space of the smart home, I engaged in a close reading of space and place, interpreting how spaces reflect meaning making around digitality and how digitality influences meaning-making processes. Finally, I deployed my training in critical rhetorical analysis to evaluate the constellation of objects and artifacts that exist within smart homes and pertain to smart homes. *Threshold* identifies the social and economic forces that modify space to intensify computational profit and alienation, often at the expense of precarious publics. As such, the book attends to the gendered, raced, and classed *spatial rhetorics* of the smart home.

THE SPATIAL RHETORICS OF THE SMART HOME

Living in digitality manifests spatially in smart homes' architecture, physical arrangements, interior design, and patterns of use that reflect technological imaginaries. Scholars suggest that "all spaces hold some kind of rhetorical charge, even if it is one of seeming neutrality."[102] Homes are spaces that appear figuratively neutral due to their quotidian nature—they are often conceived as the background against which people live their lives. Yet home itself is a site rich with meaning. Smart homes in particular are rhetorically charged since they technologically configure material and virtual spaces with selected values such as efficiency, comfort, familial relation, consumption, and production. In turn, these domesticated values are configured and reconfigured by the parties and cultural forces of digitality. If it is true, as architectural historian Sandy Isenstadt argues, that "social life literally 'takes place,'" we ought to attend to the relation between spaces, processes of placemaking, and the mediating influence of digital technology as the connective tissue holding the smart home together.[103]

Smart homes communicate the values of living in digitality spatially by infusing space with technologies and the cultural ideals of efficiency, production, and consumption. For this reason, *Threshold* extends Jessica Enoch's methodology of "spatial rhetorics" to the smart home space.[104] Enoch's "feminist spatio-rhetorical lens" contends that "spatial rhetorics are the multimodal ways through which spaces gain meaning."[105] Spatio-rhetorics "suggest the purpose of the space; the actions, behaviors, and practices that should happen inside that space; and the people who should occupy it."[106] A spatio-rhetorical lens foregrounds the way spaces meaningfully cohere over time. Indeed, as Greg Dickinson has argued, "spatial analysis demands attention to the ways everyday performances are constrained and enabled by life's material conditions; the modes by which power and resistance are arrayed across discursive and material landscapes, the means by which bodies come into contact, identities are performed, and communities are developed and dissolved."[107] Rhetorical study lends a particular focal point to the study of these everyday spaces; examining "rhetorical spatialities" focuses attention on how spaces "embody, enact, and urge values, beliefs and actions."[108] In the case of smart homes, my study of rhetorical spatialities uncovered both highly visible and mostly hidden enactments of digitality as a way of living. For instance, in several of the smart homes I visited, the homes (and their champions) loudly embodied the value of optimization—often through individual smart technologies hooked up to larger infrastructures of connectivity. But it was not often that the homes flaunted the bias haunting both the domestic spaces and technological configurations that make the smart homes possible. In these cases I relied on my training as a critic and the theoretical scaffolding of living in digitality to reveal the submerged meanings and networked effects of bias in these spaces. I am always evaluating the discursive and material conditions of digitality that are immediately apparent in the smart home and those that lie deep in the foundations of these homes and the cultural imaginaries they represent. In short, the spatial analysis of smart homes that follows attends to what appears to be present and what does not.

Spaces are subject to dispute and remaking over time, and smart homes are no exception. My approach considers how humans make sense of the smart home in contested, flexible, and hierarchical ways. This process is known as "placemaking," and it is inflected by the techno-social formations of digitality as a process and as a product.[109] As Germaine R. Halegoua writes, "making places, and the places produced, can be read as sites where the expression and documentation of human experience happen and are contested . . . they are always about the inscriptions of power, bodies, and meaning within place. The practices and people that produce spaces and places are not singular, cohesive, or complementary."[110] Instead, we ought to consider the smart home in its spatio-rhetorical multiplicities as produced by the forces of both digitality and capitalist productivity.

By attending to the lived experience of digital culture in the smart home, this book adds to scholarly conversations about rhetorics of space, technology, domesticity, and the everyday. Although living in digitality highlights how smart homes organize life around and through logics of information, instrumentalization, and control, users are not (only) passive receptacles of this cultural force. Rather, by engaging in the smart home in a routinized, bodily way, humans have an opportunity to reconfigure both domesticity and digitality as they manifest in space. Attending to the spatial rhetorics of smart homes is the first step to imagining smart homes otherwise, and its recognition of power in space obligates us to consider possibilities for placemaking.

Field Rhetorics and Sites

To attend to the spatial rhetorics of smart homes, I completed fieldwork at several smart home sites throughout the United States. This fieldwork occurred over three years, between 2017 and 2020. Engaging with the smart home firsthand—as a tourist or as a resident—has helped me give nuance to my analysis by "being present, *being there*, in the places where rhetoric does its work."[111] During such site visits, I attended smart home conferences for industry leaders, toured several smart homes across the United States, and retrofitted my own home with smart home technologies. I toured three sets of smart homes. The first smart homes I visited were model homes in the DC metro area built by residential builder Brookfield Residential. These single-family artisan-style homes were ready-made smart homes built in collaboration with Amazon. The second smart home I visited was the Duke University Home Depot Smart House in Durham, North Carolina. This smart home is a smart dormitory and "living laboratory" investigating the effects of human climate change on the environment. The third smart home I visited was the Charles Family Smart Home, part of the Imagine! Colorado network. The Charles Family Smart Home is a group home for individuals with intellectual and developmental disabilities. The Duke University and Charles Family Smart Home are experimental smart homes that seek to export their knowledge to their broader communities. Each of these case studies revealed different facets of living in digitality.

Smart Homesites and Typology

Through my three years of field research, it became clear that living in digitality is represented spatially in the smart home in several distinct ways, some of which were overlapping and some of which were conflicting. As I examined the many smart homes I toured, I developed a typology based on their rhetorical form and function. This typology shows how smart homes cohere as a category of meaning despite significant differences in design, and despite differences in how they manifest the features of living in digitality. In the chapters that follow, I describe and analyze three types of smart homes: (1) luxury, ready-made smart homes that intensify the traditional domestic ideal, "bridging" the old with the new; (2) experimental smart homes that work to undo domestic traditions and serve the public; and (3) flexible "retrofitted" DIY smart homes that residents can achieve with off-the-shelf smart devices such as Amazon Echo and Google Home. Each smart home represents one possible future for living in digitality, although some reject overtly futuristic spatial discourses (see Table 2).

Table 2. Smart Home Typologies

	Traditional/ Transitional	**Experimental/ Living Labs**	**Retrofit**
Domestic functionality	Maintain/ Intensify ideals; rejects futurism	Disrupt ideals; affirms futuristic narratives that unsettle	Flexible/Contingent
Spatial Designations	Aesthetically and architecturally traditional; modifications of domestic space minimal	Space is dictated by mission or needs of residents; modifications may be significant; likely to connect public and private spheres	Altered incrementally with introduction of technology; often maintains initial spatial configurations as technology adapts to space
Technology is figured as . . .	Ordinary, banal, or hidden	Meaningful insofar as it supports the mission of the living lab	Accessible, platformed, personalized, and customizable
Living in digitality . . .	Persists unencumbered as a luxury experience for valued digital subjects	Is challenged, in part or whole, by community-based stakeholders who seek to rework digitality as shared, public good	Is affirmed or challenged piecemeal; is propagated by the supported functionality of individual devices

TRADITIONAL/TRANSITIONAL SMART HOMES

The first set of smart homes is largely traditional in aesthetics and functionality. They seek to "build a bridge" from traditional conceptions of domesticity to smart home technologies in the present tense. These smart homes provide a gradual on-ramp for curious but hesitant smart home residents. In so doing they leave most basic aspects of domesticity unchanged but supercharge them with smart devices. Not made to be experimental, they seek to heighten the modern domestic ideal through technology. Architecturally, they blend in rather than stand out. Their primary functions are efficiency, productivity, and security.[112] Here this smart home type is represented by Brookfield Residential's ready-made, builder-quality, luxury smart homes (as described in chapter 1) . These luxury smart homes are pitched primarily to upper-middle-class professionals who wish to optimize, not upend, their domestic traditions.

In houses typical of the traditional/transitional smart home designation, smart technologies solve existing domestic problems and fit into existing domestic routines. Smart technologies augment, automate, or otherwise alter basic home tasks like cooking or cleaning. For instance, these smart homes might include smart refrigerators that monitor one's grocery supply. This smart refrigerator may be connected to the network and to other platforms, including meal planning services, grocers, and grocery delivery services. These smart refrigerators can use cameras to track ingredients. When the refrigerator registers the fridge is low on a particular item, it can prompt the user to add it to a grocery list. The refrigerator mediates the grocery-buying experience through applications such as Instacart by analyzing a user's cooking, eating, and diet habits. The smart device routes an ordinary domestic chore like meal planning, managing household items, and grocery shopping through smart technologies. The goal is not to upend a domestic routine. Nor is it to radically alter the process of acquiring household goods (at least for the user). Instead, these devices work together to streamline and therefore optimize the traditional domestic experience for users through connected technologies.

In the traditional/transitional smart home, smart technologies are figured as ordinary or banal. Often, they are hidden, embedded in the walls or appliances with which a user may already be familiar. As a result, this type of smart home does not demand radical modification of the physical space, arrangement of that space, or architectural features. However, smart technologies may encourage small, palatable changes to the domestic space. These smart homes may feature smart systems that gently modify space by introducing encoded objects into it. On occasion, the traditional/transitional smart home might make small but significant changes in the spatial arrangement of the home. For instance, the traditional/transitional exemplar in this book, Brookfield Residential Smart Homes, features a new room called "the Drop Zone." This small room was designed to receive grocery and other deliveries. It features a web of smart appliances, including smart locks, motion-activated cameras, and a closed-circuit system for monitoring the room at a distance. This drop zone integrates novel technologies and services in the home, intensifying domestic experiences through smart frameworks, but it largely fits in with existing narratives about homelife.

These types of smart homes are made to speak directly to individuals interested in dabbling in smart technologies rather than radically reworking their lives as a result of them. Futuristic narratives about smart technology are likely to be downplayed in favor of a transitional narrative. Novel technologies are normalized by literally and figuratively domesticating them, "situating the object somewhere between 'ordinary' and 'desirable'" such that the "technology must become necessary . . . while at the same time representing an improvement of the ordinary."[113] In distinction to earlier versions of the smart home, the traditional/transitional smart home purposefully avoids overtly futuristic descriptors or narratives. With radical futurism too unsettling for their ideal customer, smart technologies are presented as intuitive and seamless. The technologies are unassuming, thoughtfully configured, and coherent. The smart home traverses the rhetorical distance between technological innovation and the maintenance of familiar domestic sensibilities. As a result, these homes may be more likely to be "smart-from-the-start," or built smart from the ground up.

EXPERIMENTAL SMART HOMES (LIVING LABS)

The second set of smart homes is decidedly more experimental than traditional/transitional smart homes. These "living laboratories" question the very essence of modern domesticity. In marked distinction from the traditional/transitional smart home, these smart homes draw from the experimental ethos of earlier eras of smart housing. Experimental smart homes extend and intensify the experimental drive, using modern technologies to push the limits of homelife. Here, technology is put to service to solve a political or social problem. In so doing the living laboratory makes an explicit argument that domestic life is connected to the public sphere.

Rather than domesticating technology, these types of smart homes reexamine the traditional narratives of home. They quite literally become living laboratories where people can live and work in community with one another, and, in some instances, with the public. In this type of smart home, the domestic space is not (only) a refuge for its residents but a site of political and social change. Technology's unsettling capacities are welcome here. The living laboratory understands smart technology as useful for domestic experimentation. The technologies provide a specific mission-critical service or function, which may upset deeply ingrained paradigms in the home and beyond it. At the same time, these smart homes remain subject to the organizing logics of digitality, which limit their revolutionary capacity.

This book features two experimental exemplars. Chapter 2, "Smart Homes as Living Laboratories" examines two experimental smart homes in Durham, North Carolina, and Longmont, Colorado. The former smart home is a smart dormitory at Duke University. The goal of this smart dormitory is to find sustainable domestic solutions to battle climate change. Students who live there partner with university, community, and corporate stakeholders to advance their sustainability mission. The latter smart home, the Charles Family Smart Home, is a group home for individuals with intellectual and developmental disabilities. In this case smart technologies fill gaps in medical and other technologies, which are often very expensive and inaccessible. Both smart homes are self-described living laboratories, in which residents and other stakeholders experiment with smart technologies to support their mission. Both the smart dormitory and the Imagine! Colorado smart home partner with community leaders, institutions, and the general public to complete projects. They also seek to share their findings with broader communities through outreach programs. As seen in chapter 2, these smart homes provide a model for other smart homes by challenging the norms of domesticity.

As these exemplars evidence, living laboratories figure the home as a site for political and social action. They work to serve the publics beyond the home, as well as their residents inside of it. The architecture of these living labs is guided by the chosen *topoi* of experimentation, such as sustainability or accessibility. They are distinct from the previous type of smart home because they hope to refigure domesticity as a tool to solve major public issues rather than serve an individual family. Not only do these homes reconfigure key aspects of domestic living, but they also seek to transmit lessons learned from the smart homes to other domestic spaces. In so doing they find ways to reimagine what a smart home can be, and for whom.

Opening up the home as a site of experimentation invites new configurations of space and domestic relation. Experimental smart homes may challenge the domestic nuclear ideal, in which single-generation families related by blood live alone in stand-alone housing. Instead, householders in experimental smart homes might be drawn together by some other shared interest or affinity. These homes look more like group homes, shared housing, or dormitories than luxury single-family homes—because they are.

The function of technology in these spaces is based on the particular mission or vision of the myriad stakeholders in the experimental smart home project. Rhetorically, smart technologies are given meaning through their experimental purpose. Smart technologies are used strategically to manufacture conditions in which the home turns into a living laboratory. The mission of a living laboratory might reconfigure the space in significant ways. For science-oriented experimental homes, for instance, space may need to be modified to include actual laboratories. Living areas may need to be set off from collective work and leisure areas. Alternatively, there may need to be more bedrooms to account for privacy needs in shared housing.

These living laboratories often prioritize public engagement. In addition to challenging the domestic ideal of the home as a private space, experimental smart homes may actively encourage partnerships beyond the residents of the home. Like other types of scientific laboratories, experimental smart homes often work with external stakeholders positioned as partners in experimentation. The smart home becomes a gathering place for publics, persons organized together in community around a shared interest. These stakeholders all may or may not be living in the space. Some may live elsewhere, choosing to join the home on specific occasions. As a result, the figure of home is reconsidered as a shared site of community. The reach and membership of that community may extend outward, such that home becomes a site of engagement and gathering for people who are a part of that community—whether or not they sleep there at night. As chapter 2 reveals, these smart homes provide a model for other smart homes even as they find ways to reimagine what a smart home can be and for whom. Although these smart homes challenge the traditional domestic ideal, they remain constrained by technoliberal spatial rhetorics that prioritize profit over progress.

RETROFIT SMART HOMES

Not all smart homes are built smart. Rather, many smart homes are made smart by incorporating smart devices, speakers, hubs, sensors, and more, installed by the homeowner or another party. As such, the third type of smart home analyzed in this book is the retrofitted smart home. These smart homes are DIY versions of the smart home. Retrofit houses transform "un-smart" homes to "smart" by installing off-the-shelf technologies, including individual devices and appliances and smart home hubs equipped with artificially intelligent virtual assistants, such as Siri and Alexa. This is the most flexible and accessible smart home format.

While the third typology of "retrofitted" smart houses is distinctly spatial, its function does not rely on any particular spatial configuration. Indeed, the retrofit smart home is defined by its spatial and narrative mutability. These homes are not limited to a particular configuration of space or, indeed, "smartness." Rather, these smart homes are defined by their ability to, as Kitchin and Dodge observe, *transduce* or alter space with the incorporation of smart technologies.[114] These homes work in distinction to "smart-from-the-start-homes" mentioned in the first typology. Rather, any home can become smart if it is properly outfitted with smart technologies.

Retrofit smart homes also gesture to incremental domestic changes through technological mediation. Users can make their homes smarter by acquiring individual devices. Although sometimes sold in a set, many of these off-the-shelf devices are sold individually. As users acquire individual devices, they can calibrate their experience of living in digitality. Many smart technology makers have invested in a DIY or retrofit smart home approach. Apple, Google, Amazon, Samsung, and more are all producing and marketing smart products whose primary audience is ordinary people without any particular technological expertise. A single smart speaker, plug, or smart appliance can be easily embedded into a domestic space. Many of these stand-alone devices are affordable. The Google Home Mini can be purchased for less than twenty-five dollars; an Amazon Echo Dot costs about the same on sale. These technologies are designed to be used by nonexpert individuals and overlaid on an already-existing space. Individuals can therefore acquire a smart home by retrofitting the home they already have in a piecemeal way for relatively low start-up cost.

Of course, some individuals might prefer to develop a retrofit smart home systematically. An extension of hobbyist home automation, DIY configurations of the smart home invite users to participate in reimagining their spaces. Retrofit smart homes can create conditions under which technology enthusiasts can calibrate a space according to their needs, interests, and desires. Here, smart technology might be meaningful to users because of its technological capacities and represents an opportunity to tinker with technologies as a hobby. Retrofitting the smart home can also help users feel "in control" of their technological environments. As Gram-Hanssen and Darby note, "one way for occupants to keep control of their homes might be through DIY approaches to smart homes. . . . In the DIY approach, occupants are taking control through the technology to a much higher degree."[115] A sense of control might be important, especially for users with privacy-based concerns about the wholesale application of smart home technologies.

These three types of smart homes—traditional/transitional, experimental, and retrofit—evidence the diversity of smart homes available currently. This typology also demonstrates the extent to which different audiences approach smart homes differently. Although there are significant distinctions between each smart home type, the smart home endures as a coherent, if overdetermined, category of meaning. Rhetorically speaking, the smart home concept is a shared accomplishment, the product of diverse, multiperspectival meaning-making processes. These processes are inflected by related cultural discourses regarding the role of technology, housing, family, and public and private life. Smart homes mediate life through technologies, but the way that users understand, narrate, and engage with these technologies influences how they experience living in digitality.

BLUEPRINT OF THE BOOK

The books' argument unfolds in three parts. This introduction provides the methodological and theoretical scaffolding of *Threshold*, interweaving platform studies, rhetorical criticism, and critical theory. Chapters 1 through 3 are case studies highlighting the rhetoricity of three distinct types of smart homes: traditional/transitional, experimental, and retrofit. The conclusion envisions a smart home future unbound by technoliberal frameworks by instead drawing on feminist theories of home and care.

Chapter 1 uses the concept of infrastructure to highlight the platformization of the smart home. This chapter suggests that the cultural infrastructure of digital domesticity is literally and figuratively reshaping the domestic space. Using fieldwork gathered at smart homes throughout the District of Columbia metro area, I analyze luxury smart homes' spatial and material rhetorics with smart tech built-in. In particular, this chapter focuses on luxury smart features including a Drone Landing Pad (for deliveries) and a curious new room called the "Drop Zone," made for receiving deliveries securely. This chapter also explains how the domestic platform economy has intensified in the global COVID-19 pandemic.

Chapter 2 examines smart homes as living laboratories. This chapter draws on fieldwork completed in the summer of 2019, during public tours of two public, experimental smart home spaces. These two smart homes, Imagine! Smart Homes in Boulder, Colorado, and the Home Depot Smart Home at Duke University in Durham, North Carolina, exemplify smart homes that use private spaces to reimagine more just futures for both their inhabitants and those beyond their thresholds. This chapter shows how experimental smart homes argue for egalitarian, inclusive, and harmonious living with implications in the social, economic, and political spheres.

Chapter 3, "The New Old Housewives," attends to the artificially intelligent virtual assistants that make possible "the smart" in DIY, retrofit smart home. These devices (and their attendant personas) are simultaneously novel and familiar insofar as they are gendered domestic artifacts that purport to make life easier for their users. However, in response, they reinscribe pernicious social relations, including hegemonic gender relationships and surveillance capitalism, within the material fibers of the smart home. This chapter extends extant research on gender, capitalism, and AI VA by exploring new iterations of virtual assistants figured as the wife of the smart home. It also provides a new analysis of recent calls to de-gender devices within a larger context of gendered domesticity.

Chapter 4 asks "What can smart homes be?" It reflects on technological, social, and political infrastructures that make smart homes possible and imagines alternate pathways of smart domesticity. Bookending the case studies in chapters 1 through 3, this chapter suggests that smart homes in the present tense largely maintain dominant discourses about the home and technology, with few notable exceptions. These dominant discourses prioritize capital and data accumulation over care and wellbeing; in so doing, the smart home subverts the most radical aspects of domesticity for profit. Mobilizing bell hooks's concept of "homeplace," the chapter highlights the political nature of the smart home as a site of "liberatory struggle."[116] This conclusion chapter affirms that the smart home is an interface for the technological future because it is the space where discourses about the future are enacted. Extending the smart home concept into the future, this chapter incorporates demands from technology and domesticity scholars and activists to reimagine the smart home as a site for radical change.

Together, these chapters highlight how smart homes' use of digital technologies (re)configure space, affirm and reject domestic sensibilities, and make an argument for living in digitality. Using extensive fieldwork, critical analysis, and interpretive methods, I show how smart homes expand the vertiginous reach of the platform using familiar, and often troubling, domestic narratives. *Threshold* tells the story of smart homes as they have been in the past, are in the present, and how they might be in the future.

1

Luxury Smart Homes and the Stranger Other

In a quiet suburb of the District of Columbia, homeowners lounge on plush furniture, sip organic herbal tea and order the next week's groceries from an app on their phone. Later, while they work in the office or luxuriate at home, a stranger will drive up to the two-story "artisan-style" house. Without disturbing the homeowner, this person will use a passcode to enter a special side room. The space is one of the first of its kind—made to respond to the new needs of a modern homeowner. The stranger will deliver groceries there, expertly placing the requisite goods into the stainless-steel refrigerator provided for this purpose. Later, the homeowner will transfer these goods to a second refrigerator in the spacious, well-outfitted kitchen. And just as quietly and efficiently as they came, the delivery person will leave the house, rushing to fulfill the next order on a tight timetable. The homeowner will have a record of this stranger's visit, made possible by several cameras, sensors, and other devices connected to the internet and hardwired into the home. The stranger will too, but it will show this house as one of many stops managed algorithmically by a platform. The interaction between homeowner and stranger will be minimal, and it will be mediated through a technological and economic infrastructure that organizes their relationship with each other. This is the domestic future promised to a new class of smart home owners in the affluent suburbs of DC and throughout the US.

This chapter traces the spatial expression of living in digitality in built-to-suit homes in the Washington, DC, area. In particular, I investigate Amazon's partnership with Brookfield Residential (BR), a major residential developer. Together with Amazon, BR has developed a new product, the smart home. The BR Smart Home offers customizable, luxury smart living, including the traditional (stainless steel appliances, a butler's pantry) and the transitional (an expansive security system, a drone helipad, a new "drop zone" [DZ] room for Amazon and other deliveries). An exemplar of mass-market smart home spaces proliferating throughout the United States, the BR Smart Home represents one possible future of living in digitality.

In the following paragraphs, I offer readers a tour of the BR space, with a particular focus on the home's "smartest" features. These features include smart home products like devices, appliances, and sensors. The BR Smart Home also advertises as a feature new smart spaces that align with the home's broader smart ecosystem. The BR Smart Home sees the *interior world* of domestic space incorporated and optimized within the logics of digitality. The features of the smart home also help to configure the *outside world* as inherently optimizable, extending digitality toward the public sphere. For instance, the BR Smart Home features a built-in ambient surveillance system, which automates the experience of risk management between the household and the outside world through connected technology. Smart appliances in the BR Smart Home "speak" to one another, routing domestic data through an ever-expansive platform that characterizes, quantifies, and administers life-in-advance through the platform's infrastructure. Other smart features like delivery drop zones spatially segment populations into self and stranger while regulating the domestic sphere's relationship to work, care, and well-being. Luxury smart homes like the BR Smart Home mobilize an increasingly contingent labor force in order to execute the smart home's most fundamental functions. The BR Smart Home connects the domestic sphere to the platform economy, producing and affirming unequal domestic relations across multiple scales.

My argument in this chapter is that the BR Smart Home is a mass-market expression of *domestic futurity*, which I define as a cultural condition of living in digitality in which the home space is increasingly mediated by platforms, technological products, informational logics, and what Lynn Spigel calls "ambivalent discourses" about a domestic future that may come if/when we deploy smart technologies.[1] Domestic futurity is expansive, immersive, and productive. It provides a narrative and spatial scaffold to imagine the next smart home future. In other words, domestic futurity is a form of *cultural infrastructure* made possible by the expansive and networked infrastructure of digitality. Domestic futurity's cultural infrastructure helps to architect relationships between strangers inside and outside of the home. As an infrastructure, domestic futurity is a profoundly *environmental* and *agential* force, using the logics of digitality to stabilize socioeconomic relations of otherness in the physical arrangements of the smart home space. Here, domestic futurity *transforms* and *enacts*, shaping the domestic space according to the computational logics of incorporation, optimization, and bias. These smart home spaces, which are given meaning by the people who live and work within them, reflect and reaffirm the logics of digitality by encouraging domestic habits within these computational paradigms of smartness. Domestic futurity relies upon and reifies larger cultural trends, in particular, the widening gulf between smart home owners and stranger-others, especially through the performance of what some scholars call "low-income service labor" in the platform economy.[2]

Like digitality, domestic futurity is characterized by structural hierarchy and difference coded into the machine from the ground up. As a result, luxury smart homes assist in the calcification of difference inside and outside the home. For homeowners, the luxury smart home purports to offer a carefree existence through the wonders of automation and algorithmic processing. For all others implicated in the running of the household, smart devices offer a different technological future. When domestic futurity is the dominant framework for smart home imaginaries, the sociotechnical arrangement of smart home spaces figures the stranger-other as a necessary but risky resource for maintaining a luxury lifestyle. Smart devices and systems are pressed into service not only to execute daily tasks or increase leisure time but to manage this everyday risk. Instead of experiencing ease and care, individuals outside the smart home are understood as fungible and subsequently subject to various forms of algorithmic management.

This case study furthers our understanding of the politics of infrastructure by showing how cultural forces such as domesticity act as infrastructure and shape the human experience within and beyond the domestic sphere. The BR Smart Home also demonstrates how myriad infrastructures converge in the home, a space often figured as discrete, individual, and private rather than structural, collective, and public. This chapter aims to reveal how, left unchecked, this form of living in digitality can and will manifest as an inhospitable sociotechnical condition for all but the wealthiest digital subjects. The case study suggests how living in digitality manifests in an idealistic form of smart home living in which apparent optimization logics subsequently embed bias and technical control into digital spaces. Importantly, the smart home exemplars that I investigate forward a form of exclusionary domesticity that is couched in the reaffirmation of a bourgeois and hierarchical domestic ideal, reminiscent of earlier exclusionary domestic logics. To this ideal, digitality adds a mediating structure, merging together emerging technologies and political economies that orient smartness around luxury for few and precarity for most. The result is an overly corporatized smart home product that magnifies the effects of the inequitable systems on which digitality runs.

DOMESTIC FUTURITY AS INFRASTRUCTURE

The chapter advances two overarching claims about the infrastructural nature of smart homes. First, domestic futurity is infrastructural insofar as it provides a spatial and ideological architecture for politics, technology, and culture in smart home spaces. The infrastructural components of domesticity shape the individual spaces and collective narratives of home while also producing subjects that are "assigned socially acceptable roles that are supported by the architecture of the house."[3] As I claim elsewhere in this book, home is both a physical place and collection of ideals wherein cultural actors negotiate public issues, including issues pertaining to race, gender/sex roles, familial relations, autonomy, and labor.[4] Domesticity is both the product of ideological forces and itself a force for managing power relations inside and outside the home. For instance, bourgeois, middle-class imaginaries about the role of the domestic space persist as idealized standards against which to live, despite significant challenges to those idealized standards inside and outside of the home. Often these hermeneutics of power and technology are operationalized in space, such that the home becomes a site for enacting unequal relations in the domestic interior and without. Idealized expectations of the domestic sphere are both local and distributed; networked technology helps manage the relationship between the domestic and the public sphere and is mediated by existing ideologies. At the same time, the rhetoricity of domesticity provides a coherent and value-rich narrative to justify otherwise dated domestic relations.

Homes are also positioned in dominant cultural imaginaries as private havens to develop and practice the "good life." These conflicting paradigms situate the home as a useful and productive cultural space that should be managed and secured from outside threats. In the BR Smart Home, technocultural narratives about the sanctity of the home structure the physical footprint of the domestic space as well as expectations for what humans do within/beyond it.[5] The BR Smart Home exemplifies and spatializes the multiscalar, expansive, and biased condition of digitality in theory and in practice.

Second, all smart homes are always already networked infrastructures that represent the logistical interconnection of macrosystems (such as capital flows) and microsystems (such as the family unit). As a result, smart homes serve as the platform where domestic commodities endemic to an industrial post-Fordist age merge with a service-based economy to produce a networked domestic infrastructure. Although characterized by innovation, the function of the BR Smart Home is ultimately regressive. As Spigel argues, "the home of tomorrow functions most typically as a deeply conservative structure that promises a version of technological progress based on nostalgic longings for privacy, property, and propriety."[6] In the case of the BR Smart Home, ubiquitous platform technologies aid in the production of domestic futurity wherein these nostalgic longings are spatially rendered in rooms configured to incorporate stranger-others while also securing the home against them. Here, the needs and wants of privileged domestic subjects are prioritized over the economic and physical safety of the gendered, classed, and racialized stranger-others who labor to make the Smart Home possible.

The BR Smart Home is figured by the broader political economy of the digital domestic sphere, including surveillance capitalism and platform labor.[7] In the case of the BR Smart home, luxury digitality reorients the smart home space as both inside and outside the economic sphere. Through the connected capacities of digitality, the smart home space is protected from the external world but also rendered productive because of its carefully mediated relationship to it. In this smart home configuration, living in digitality is characterized by commercialized rhetorics that promise ease, efficiency, and technological optimization for homeowners while also intensifying existing inhospitable hierarchical social relations within and beyond the connected home. The BR Smart Home evidences that domestic futurity is defined by persistent ideological discourses organized around the feature of a stranger-other who exists and labors within and beyond the home's walls.

BR: THE FUTURE, TODAY

"This is how we're going to live. All of us." This phrase serves as a repeated refrain in advertising for BR Smart Homes. BR's Smart Homes are a recently developed smart home product that pairs the promise of living in digitality with the ease and efficiency of a new domestic build. These smart homes communicate the values and (infra)structural function of domestic futurity in several ways, including through the medium of several open-house, tourable smart home models spread throughout the suburbs of the District of Columbia. In the summer of 2018, I toured most of them. Although each smart home model had a different floor plan and interior design motif, the promise of luxury domestic futurity was a repeated theme throughout all tours. Many of these luxury smart homes feature mass-market technology that automates domestic life, including cooking, cleaning, grocery shopping, and more.

BR Smart Home models communicate ideals of domestic futurity in luxury homes and townhomes. Like many residential builders, Brookfield develops model units in strategic neighborhoods to advertise their basic housing products and innovative smart home collaborations. The neighborhoods in which the smart homes were (being) built were freshly (re)developed areas subject to gentrification processes as they were made ready for living in digitality.[8] The majority of the homes were in predominantly white, upper-middle-class areas.[9] The median household income was well above one hundred thousand dollars a year.

In comparison, the American Community Survey estimates that Black or African American populations constitute some 47 percent of the District of Columbia proper, with white populations making up around 40 percent.[10]

The median household income in the District is around $77,000 per annum.[11] These numbers suggest white flight from DC to its more affluent suburbs. Indeed, several smart home communities are characterized by "booming development," resulting in the ouster of poorer and minority families, who "must look for less expensive homes in older suburban areas," effectively segregating some schools.[12] Smart homes can contribute to gentrification of areas both inside city centers and in more suburban locales. Such gentrification further concentrates wealth and power in the hands of the few to the express detriment of the many. By ousting poorer communities, or communities of color, the "booming development" of smart home products cements living in digitality as an exclusive and exclusionary condition. Importantly, the narrative of "development" and "convenience" is associated with domestic ideals separating privileged digital subjects from the masses. For instance, many of these neighborhoods have access to major transportation infrastructure, including the Metro, presumably enabling their inhabitants to commute from the suburbs to work in the city. Even so, nearly every newly built home I saw had a garage. Access to public amenities is often touted by those "developing" areas for these privileged subjects. The ability to bypass the public is a similar, if not less discussed, amenity.

I toured three BR Smart Home models, in Bristow, Virginia; Prince William County, Virginia; and North Potomac, Maryland. During my visits, I was offered tours of various lengths by real estate agents and others employed with BR. These tours frequently began in an office near or connected to the smart homes. There, I introduced myself as a university researcher of smart homes. Before I entered the homes, I was offered access to promotional material about the neighborhoods, the smart home, and Brookfield. These promotional materials primarily highlighted Brookfield Residential neighborhoods, features, and their various model options. However, pamphlets and flyers were also on display about the benefits of a particular neighborhood or township. After a guided tour, I was allowed to tour the residences with my photographer privately. Together, we explored the routes offered by tour guides and, when left to our own devices, deviated from those initial routes to explore the home.

Under a corporate partnership between BR and Amazon, a majority of the smart home features are couched in commercial rhetorics. Corporate "technologist" communication about smart homes is a constitutive component of living in digitality, in part because "smart homes are a particular sociotechnical vision developed by technologists . . . promulgated by a nexus of product designers, house builders, and appliance manufacturers, and increasingly the software industry, focused on driving new rounds of consumer fashions and home upgrades."[13] These commercialized rhetorics about living in digitality manifest in overt and covert ways. Here, domestic futurity is signaled as an accompolishment of partnerships between individual homeowners and platforms, evidenced by gratuitous signage throughout the home noting the Amazon-Brookfield association. Indeed, Amazon-branded products (or those associated with Amazon) constituted a great deal of the smart technology throughout the home. Because "architecture has become an instrument in the commercialized construction of experience," the homes' material configurations support the experience and expression of commercialized domestic futurity.[14] Analyzing smart homes as a medium for a particular form of infrastructure reveals both the traditional and transitional facets of sociotechnoculture(s) in the present tense.

Avendale

In offering a brief glimpse into luxury smart home products, I leave unremarked several remarkable features of the BR Smart Home. For instance, I choose not to focus on the bevy of slick, stainless-steel smart appliances. I also choose not to ruminate upon the specific staging of rooms crafted by smart home technologies throughout the BR Smart Home. For instance, I decided against a protracted reading of the character-themed (styled as a boy's) bedroom and the pop-star disco-party-themed (styled as a girl's) bedroom. These smart features, while interesting, are beyond the scope of this chapter. Instead, I chose to focus my interpretive efforts on the "smartest" parts of the home, those spaces in which domestic futurity indelibly altered the footprint of the house.

I toured three distinct houses in three different neighborhoods and townships, but it became clear that the first house I toured was the most robust in terms of domestic infrastructure. The Avendale house boasted the most interior smart upgrades and was the most clearly marked as a smart home. The other home models I toured largely replicated Avendale's features. Thus, for this analysis, I am choosing to focus on this smart home in particular.

After meeting my tour guide in an adjacent house-cum-sales office, our Avendale smart home model tour began outside. The two-story "artisan-style" single-family home was anchored in a verdant, well-manicured lawn (Figure 1). The home's front elevation was clad in a formidable but welcoming dusty red brick. Crisp, white-trimmed windows with black shutters revealed glimpses of the home inside. A small porch covered the front door and one of the windows. To the far right of the front elevation stood an attached two-car garage. The smart home commanded the corner lot on Twin Leaf Drive, backed up to tall, skinny, deep green trees gently swaying in the summer sun.

My view of the front elevation was partially obstructed by a thin, black, railed fence, dividing the front yard somewhat unnaturally in two. The railing created a semi-fixed corridor linking the smart home model to the BR office housed (literally) next door. There, a BR agent may offer branded literature before guiding an interested client on a predetermined path throughout the smart home. In the Avendale office, large posters highlighted the prestige of the neighborhood and the functionality of Brookfield's new builds. The visitor was invited to "Save Energy and Money" with included energy-efficient features. These facets of domestic futurity are carefully patterned luxury constructs; as another large poster noted, Avendale is among "The Best Places to Call Home," because of "convenience, access, and a real connection to the surrounding community." A small sign set in front of these two posters listed "Luxury Single Family Home Pricing," with most options starting around half a million dollars. These advertisements primed the visitor to experience domestic futurity through the lens of luxury, convenience, community, and efficiency—repeated themes in the space.

This home looked nearly identical to other homes lining the block. The front elevation read traditional American suburbia rather than domestic futurity to my eye. Having studied "Yesterday's Houses of Tomorrow," replete with experimental architecture, I was surprised by the traditional structure that stood before me.[15] In 1962 folks experimenting with domesticity proclaimed, "It's not likely that space-age man will retain the prison of today's straightline architecture."[16] Prison or not, the straightline architecture persists. The invisibility of smart architecture is, in many ways, the defining feature of mass-market smart domesticity, which figures innovation through tradition.

Figure 1. Avendale Smart Home. Courtesy of James Alexander McVey.

The Drone Landing Pad

Following the black fence transecting the front yard, my friendly, knowledgeable, and energetic BR agent led me to the smart home. Perhaps sensing that the smart home's brains were hidden from me, my tour guide showed me arguably the most technologically advanced feature of the home first. There, beside a side-door entrance, discreetly tucked away from the front elevation, is a raised, square platform set atop stone pavers. On it, a large circled B is painted in blue. A sign next to the raised platform indicates the utility of the feature: "the smart home: Drone Landing Pad" (DLP, Figure 2). My tour guide explains that the DLP is made to receive Amazon package deliveries shuttled via drone. I scan the skies, imagining the logistics of drone package deliveries in this suburban neighborhood. None appear.

Although neither Amazon's name nor any apparent branding appears on the DLP, Brookfield's Smart Home partnership with Amazon enthymematically gestures toward Amazon's "Prime Air" delivery service. Prime Air is a parcel delivery service in development at Amazon, which promises to deliver packages "in 30 minutes or less using unmanned aerial vehicles, also called drones."[17] The plan for Prime Air delivery was announced in 2013, but critics noted that the service was likely untenable, for regulatory and technological reasons.[18]

Figure 2. Drone Landing Pad. Courtesy of James Alexander McVey.

An important feature of Amazon's Prime subscription service is "free" two-day delivery for many consumers. As the service has grown, Amazon has experienced logistical issues related to delivery. One is a dearth of delivery personnel, including from third-party contracted services not intimately associated with Amazon. Platformed logistics solutions like Amazon Flex pay people—deemed independent contractors—to deliver Amazon packages using their own vehicles and smartphones. Commentators nevertheless argued that "there aren't enough drivers to keep up with your delivery lifestyle."[19] Amazon effectively sold the promise of rapid delivery service without first scaling up a shipping infrastructure to make those delivery times possible in the long term. Concurrently, Amazon has been criticized for labor rights violations, including poor working conditions for factory and warehouse workers and delivery persons who make rapid turnaround possible.[20]

Amazon has experimented with several platform technologies to increase efficiency on its retail side, many of which are invisible to the end consumer. These efficiency measures transform invisible stranger-others into logistical components of the warehouse. In 2016 Amazon filed a patent for a surveillant wristband that would track and quantify workers' productivity and offer suggestions on how to be more productive.[21] Amazon is also exploring alternate, nonhuman routes to meet consumer demand. "Prime Air" is one example of Amazon's experimentation. At the BR Smart Home, the DLP appears a bit worn, although apparently not from drone delivery use. Prime Air's "rapid parcel delivery" successfully delivered a test package on December 7, 2016, nearly three years after Amazon pitched the delivery service.[22] Since then, Prime Air has largely been grounded. In 2019 Amazon issued a press release announcing a new drone infrastructure. Touting advances in technology and safety, Amazon suggested it could begin drone delivery "within months."[23] Whether or not the drones are a stable delivery technology, they represent the desire for delivery without people.

The DLP is consistent with what some scholars have called the "Uberization" or "Uberification" of the economy.[24] The Uberification of the economy is characterized by algorithmically managed connections between individuals and products or services. Platforms organize and orient this connection, transforming traditional marketplace processes by offering consumers on-demand access to goods provided by (or through) individuals who work contingently. In theory the platform cuts out the "middle man," but in practice the "middle man" is not only very present but is also marginalized by the technical logics of the so-called sharing or peer economy. Uber drivers offer not only their time, emotional labor, and skill but their own infrastructure to move goods. These "independent contractors" are unlikely to be materially associated with either the organizing platform or the proprietor offering the good or service. Rather, they exist in the margins of a new platform economy that relies on their labor but does not recognize or fairly compensate it. The DLP represents the fullest expression of the Uberification of the economy by removing the humans already understood as redundant. Whether or not drone delivery works does not matter; what matters is that more "efficiencies" can be slotted into an already "optimized" market.

The DLP serves a symbolic function, a gesture to a domestic futurity mediated by Amazon and partners. The DLP hints that what makes the smart home smart is the invisibility of supplier-side logistics. The invisibility of stranger-others (and their labor) may obfuscate larger structural hierarchies that constitute economic and social flows and Amazon's retail logistics specifically. As Burrington notes, Amazon is exceptionally deft at "abstracting infrastructural problems into logistics problems" and "away [from] a lot of the more discomforting or difficult facets of its infrastructure," including poor working conditions for its employees and affiliates.[25]

In the context of infrastructures, the DLP supports a domestic futurity that is abstracted within and beyond the home. Domestic labor required by smart home owners need not be done by the humans residing in the home or by humans at all. By implicitly gesturing toward the removal of the human, the DLP affirms the invisibility of stranger-others who support luxury domestic futurity. The benefit of such a smart technology is that it equips the homeowners to harness the power of Amazon's logistics infrastructure. This flashy display of innovation may be persuasive enough to obfuscate poor labor conditions on the other side of the drone controls. At BR, the DLP is a primary feature defining the home's smartness as made possible by platform-mediated domesticity. As the first official stop on the smart home tour, it holds significance as the preeminent example of domestic futurity. The next stop on the tour would continue this trajectory while highlighting features that erect barriers between the homeowners and stranger-others within the smart home.

The Drop Zone

The Smart Home's expression of a corporate, platform-mediated domestic futurity extends once we enter the side doorway of the home. To the left of the door is a keyless entry keypad and what appears to be an intercom device. As I turn from the DLP to enter the door, I notice a wall-mounted camera affixed above the outdoor lighting sconce. I immediately wonder who is watching me, and who has a record of my visit to the smart home. The utility of these surveillant devices is made clearer to me as I move through the door's threshold into a type of room I have never seen before.

Just inside the doorway is a small room—so small that it barely fits me, my photographer, and the tour guide. We pack inside the novel, clean space equipped with a stainless-steel refrigerator, crisp white cabinetry with metallic finishes, and a small, white Amazon Echo Dot, a disclike smart device. Next to it, on a granite-lined countertop, sits a woven picnic basket and an apothecary jar decorated with small, wax red and green apples. Immediately above these decorative items is a white-framed picture window. This window is a protected portal to the home's interior. Although an inviting living room sofa is in sight, it is unclear to me how to access it.

The space is furnished with a blue and white area rug and wall decor. A decorative sign above the window advertises "Butter and Eggs" in a faux-distressed "farmhouse-chic" font. Similar signs read "Organic" and "Fresh Herbs." Although I see nothing of either's referent, or any real food, for that matter, these decorative items and the inclusion of a refrigerator clarify that this room is meant to hold or prepare foodstuffs—and, based on the decor, preferably foodstuffs of organic origin—straight off the farm. This is decidedly not a kitchen, though, owing to an absence of appliances or preparatory tools. Later in the tour, I will be introduced to a full, luxury chef's kitchen. Circular signs affixed to walls tell me we are in the "drop zone," a new kind of room separated by at least two locked doors from the rest of the house. In the corner of the small room, a pile of boxes with blue Amazon delivery packaging suggests that this room's primary function is to collect deliveries. In addition to Amazon-branded materials, the decor and kitchen appliances suggest this room is made for grocery delivery service through platformed services like Instacart. This room, then, gestures to the traditional (farmhouse, organic produce) to mediate potential anxieties about the transitional (platformed grocery and delivery service that invites stranger-others into one's home). Here, the paradox of domestic futurity of the BR Smart Home is apparent; the life-changing nature of technology is both very much on display and meant to be hidden by a nostalgic longing for an agrarian past.

This room serves as a space for delivery persons to drop off homeowners' packages without formal interaction from the home's occupants. The keypad outside the door, the bubble camera affixed to the external wall siding, and the intercom all function as a way to afford remote entry for stranger-others who service the home through on-demand platforms. The home's occupants need not be present when packages are delivered. When prompted Alexa gives the DZ pitch: "Imagine your fresh deliveries being securely placed right in your refrigerator, freezer, or here, out of harm's way. You quickly grant or deny access to visitors through your smart home while you are away. With video monitoring, watch as your deliveries arrive, and rest assured your home remains secure thereafter." Should the drone delivery pad outside fail to meet the remote delivery needs of the homeowners, the DZ is equipped to handle deliveries from strangers hired to deliver their goods. The fridge also allows a grocery delivery person to keep foods fresh without needing to access the kitchen and, therefore, the home's interior. Defined by its function, the DZ creates a distinct work zone for delivery persons who need access to the home but not its most intimate interiors. The apparent need to securitize the space, paired with on-demand platforms that manage workers through contingent, short-term "gigs," rather than "jobs," suggests that delivery persons are unlikely to be known to the homeowners. Again, a paradox: to secure their purchased goods (a transitional problem), the homeowners must relinquish some access to the private, secured space (a traditional problem) to strangers.

The DZ is an expression of living in digitality. Most generally, the purported function of the smart home is to engender a life of efficiency, leisure, comfort, security, and more. Yolande Strengers and colleagues qualify these technological features as the "three Ps"; smart homes provide *protection*, *productivity*, and *pleasure*."[26] The three Ps may be experienced throughout the smart home. The DZ is a unique room where all three Ps may be experienced simultaneously.

First, *protection*. The purported function of the DZ is to protect the home's inhabitants from outsiders, including those contracted to work in the home. Like many modern workspaces, this room is secured and monitored, subject to surveillance technology affixed around and within it.[27] The DZ is monitored by smart technologies installed both in the delivery room proper and in the main interior portion of the home. In addition to its locking doors, the room is equipped with a smart lockbox and a doorbell camera. Access to this secured exterior room may be given from the interior of the house or from a distance. If the residents are home, the DZ makes a delivery without human contact possible—perhaps even probable. The main interior is further secured from the DZ and

Figure 3. Securitizing the Home. Courtesy of James Alexander McVey.

its (temporary) inhabitants. The path from the DZ to the home's interior is controlled by a bevy of networked security technologies, including two locking doors. A flat-screen television security display is near the chef's kitchen (Figure 3). This display is toggled into quadrants. Each of the quadrants reveals closed-camera security footage of the house. Notably, the DLP and the DZ each gets its own quadrant. The parts of the home subject to engagement with delivery persons constitute half of this particular surveillance command center, thus signifying the importance of the space and of securing it.

The delivered items are also subject to intense securitization.[28] Like other forms of surveillance and securitization, the persistent monitoring of the DZ is a "performative and participatory act."[29] In constructing the stage of the performance, threats must be identified and ways to foil the threats highlighted. The DZ gestures toward the problem of "porch pirates," stranger-others who allegedly take unmonitored delivered packages from stoops, landings, and porches. The *New York Times* has reported on the increased police presence in response to porch pirates, including police officers conducting "sting operations" in strategic neighborhoods.[30] The threat of porch pirates has inspired new iterations of neighborhood watches. In addition to drawing upon the power of the police, some neighborhood watchers seek vigilante justice, with persons harnessing networked surveillant devices to prevent or punish package thieves.[31] The DZ and its surveillant accoutrement seem to affirm the seriousness of parcel theft as a public (and private) safety issue. The DZ assumes in advance the threat of *any* stranger-others on or near the homeowner's property, which must be secured.

Empirically, networked platforms deployed by homeowners and other residents to secure their neighborhoods have been used to profile people of color.[32] In the case of predominantly white, affluent neighborhoods (including BR Smart Home locations), nearly any person of color may be rendered suspect through user-installed, user-managed technical surveillance. Some surveillance platforms work in close partnership with police authorities. Footage from Amazon's Ring, a doorbell with a security camera and connected neighborhood watch app, may be obtained by police without a warrant, so long as the resident agrees.[33] Here, the choices of individual smart home users are prioritized, rendering the agency and choice of those beyond the home as secondary.

When smart home residents opt in to using devices like Amazon Ring under the guise of individual, familial domestic security, and "peace of mind," they are participating in a racialized skepticism of the outsider. The fear of a stranger outsider is a product of suburban white panic codified by both legal processes (like redlining) and societal norms around propriety. Katie Lambright has argued that "the restrictive housing ordinances of the midcentury contribute to the racial segregation still reverberating in the present," which in turn "fed into fear and distrust of the 'outsider.'" The outsider trope functions as a type of ideograph—specific enough to warrant preventative action but broad enough to code anyone as a threat.[34] Battling the outsider remains a process of determining interior from exterior, self from other. From at least the middle century on, the domestic space was a site to be managed, securitized, and even purified against amorphous racial threats. In other words, Ring is a technological response to the constructed social problem of the "outsider."[35] It relies on a vast technical and ideological infrastructure that conjures up (an existing) fear of the other to sell products for the "safety" of a community, with or without the explicit permission (or even knowledge) of individuals within that community.

Moreover, the domestic space extended beyond individual houses, producing a (supposed) need to countermand threats at the level of neighborhood or community. Resulting actions to secure the increasingly expansive domestic space link the neighborhood watch of the past to neighborhood social networks like Nextdoor. Such efforts are part of a broader social history normalizing citizen policing of other citizens and, given the racialized history of housing in the United State, activating individual prejudice toward a threat it has created. The BR Smart Home is nestled within and reinforces ongoing efforts to police the neighborhood and home as sanctified spaces. Such policing practices are especially problematic for contingent workers who are not likely to be known to the homeowners. This danger is compounded by the fact that "today's low-income and supposedly 'low-skill' workforce consists of people of color, many of them immigrant men and women."[36] These laborers are statistically unlikely to live in the same neighborhoods where they deliver goods. Delivery persons and other stranger-others who engage with the home are themselves securitized; by virtue of the repetitive security measures, they are also positioned as potential threats to be neutralized—especially if they are people of color entering homes in predominantly white, upper-middle-class communities. The DZ places literal walls between homeowners and delivery persons. Nevertheless, as Low argues of gated communities, "there is fear even behind the walls."[37]

By configuring a few spaces to interface with the outside world, installing some cameras and smart locks, and outsourcing logistical considerations to a platform, the BR Smart Home imagines domesticity as a hermetically sealed space that also lacks for nothing. In the BR Smart Home, strangers are figured as temporary assets in service of luxury domesticity. They are managed as such, with entire technical infrastructures dedicated to tracking the stranger-others' movements in space and time. Stranger-others who are not directly contributing to the luxury experience of the smart household are also subject to the BR Smart Home's technical infrastructures. These include closed-circuit television, networked security cameras, wireless locks, automatic doors, embedded sensors, platform interfaces, social media networks, real-time mapping products, etc.

Second, the DZ gestures to the smart home's stated promise for increased household *productivity*, even (or especially) if the responsibility of productive labor simply shifts to contingent laborers or machines. What is being produced is a protected, luxury suburban lifestyle. One important function of the DZ is to relieve occupants of labor associated with consumption. Smart homeowners may receive packages from nearly anywhere. At the same time, the addition of the refrigerator in the DZ makes the space especially useful for utilizing home grocery delivery services. The ability to order and receive groceries online is a burgeoning component of e-commerce, especially during and after the COVID-19 pandemic.[38] However, fresh, perishable foods are subject to spoilage before, upon, or shortly after rapid delivery. One of the biggest reasons for e-grocery food spoilage is that consumers cannot attend the delivery. This problem has a name: attended home delivery problem, or AHDP.[39] Smart domesticity offers solutions to AHDP. First, the DZ replaces flawed unattended delivery solutions, including delivery boxes with an entire climate-controlled room and requisite appliances for reception and storage.[40] Second, the DZ increases efficiency by making this space accessible to anticipated strangers with either a passcode or remotely controlled locking doors. Third, the DZ's security cameras allow homeowners to control this space from a distance without otherwise disrupting ongoing activities. One can easily imagine a smart home resident at work pausing to allow a delivery person entry to the DZ as part of a productive workday. Or, just as likely, remote access may be monitored and granted via smartphone from a resident tending to dinner or the children inside the home. The DZ allows for increased efficiency of consumption practices while elegantly solving a logistics issue well known to the e-commerce community.

The DZ may also contribute to a deep sense of *pleasure* for homeowners. First, ordering goods through a platform infrastructure and receiving them at one's leisure may be a pleasurable experience in and of itself. Platforms such as Amazon pair the pleasure of shopping with the ease of ordering remotely, anytime, anywhere. Second, this order and delivery schema may reduce the time spent acquiring home goods. Using a grocery-buying platform to plan and purchase meals and have them delivered may be a tenfold reduction in time spent by the consumer. This may free the smart home resident to engage in other, more pleasurable activities or—sometimes equally pleasurable—more work. Finally, as Strengers notes, early adopters of smart technologies may gain pleasure from the device's novelty—being the first to experience a smart home experience. These individuals may experience pleasure from sharing their newfound lifestyle with others.[41]

The DZ is a room built in anticipation of both known and unknown strangers, providing a service to make homeowners' lives more productive and pleasurable. Its incorporation into the smart home makes possible an increased level of security for homeowners. Of course, this protection, productivity, and pleasure come at a cost. People, platforms, and digital objects performing domestic labor are on the other side of the two dead-bolted doors. Moreover, the DZ is managed by a securitized system rendering delivery persons (or, more acutely, *any persons*) subject to scrutiny, suspicion, or even violence at the hands of the neighborhood watch. The branded materials in the DZ suggest that the people delivering goods are likely to be noncontractual, contingent laborers. Both Amazon and related delivery services such as Instacart deploy a workforce made increasingly of independent operators, who often perform what Niels van Doorn calls "low-income service labor." Spurred by an on-demand consumer economy in which consumers increasingly grow used to rapid delivery, these workers routinely labor at the margins, without job security, benefits, or appropriate compensation.[42] The BR Smart Home platform logics may also put them at physical or emotional risk, especially if they are marginalized persons already subject to high suspicion and violence. These occupational hazards are one effect of domestic futurity's infrastructure, which is realized in BR Smart Homes.

The DZ's primary contribution to domesticity futurity is what van Doorn calls "platform labor," which is inherently "gendered and racialized." Van Doorn defines platform labor as "digitally mediated service work associated with the 'on-demand' or 'gig' economies."[43] For van Doorn, platform labor as used by Uber and others serves an "infrastructural" purpose in both the labor market specifically and in the platform economy in general. Platform labor is logistically organized by "platform labor intermediaries" who help direct, control, and process contingent laborers.[44] There are a number of networked effects of platform labor. When platform intermediaries manage workers through "on-demand platforms," they "enforce the *immunity* of both the buyer of a service . . . and the company that owns the platform."[45] The effect is that platform intermediaries distance themselves from workers who cannot claim worker protections and regular employee benefits when they are designated individual contractors. Second, platform intermediaries also control contingent workers by closely monitoring their performance and algorithmically organizing labor according to predictive analytics.[46] Third, platform intermediaries render service workers *fungible* or expendable, subject to replacement by new workers hired to take the place of those who burn out or cannot meet the demands of the platform or client.[47]

From a platform labor perspective, the DZ is a semiprivate domestic antechamber meant for a new class of contingent workers. BR's DZ and the DLP create a novel, semiprivate workspace for platform labor intermediaries to manage contingent laborers. In the case of BR's Smart Home, Amazon serves the role of platform labor intermediary. In this role the platform and its leadership perform a number of functions critical to maintaining a contingent and low-income workforce. In the DZ and throughout the home, contingent workers managed through the Amazon platform are subject to surveillance and algorithmic control through workforce-monitoring technology. If found wanting, their performance may be monitored and even replaced by other contingent laborers. Finally, the BR Smart Home renders laborers both hypervisible through closed-circuit television and invisible under their fungibility and stranger status. In fact, by virtue of these first two stops on the Smart Home tour, BR and Amazon seem to have a contingency plan for the eventual disappearance of on-demand human labor: drone technology that can get the job done without (many) humans employed.

The securitization of the home through the DZ is justified through traditional conceptions of privacy routed through transitional smart home technologies. Platform logics associated with the smart home determine access to varying degrees of privacy in the home. Privacy and security are afforded to homeowners, who control the visibility of an ever-contingent working class—and everyone else outside the home. The DZ space narrates the role and status of all parties involved: protected inside the home are the homeowners, who receive goods brought to them by contingent laborers. These stranger-others are paradoxically subject to intense surveillance as well as the invisibility characteristic to the platform economy.

LUXURY SMART HOMES AND COVID-19

Thus far, this chapter has examined how BR Smart Home products mediate technocultural transitions by appealing to traditional conceptions of domesticity, including security, privacy, and familial relation. In particular, I have analyzed two new domestic features mediated by platform technologies: the drone landing pad and the drop zone. These platform technologies mediate domestic futurity wherein the BR Smart Home is productive, protected, and pleasurable.

Domestic futurity manifests *technologically* through the incorporation of smart products and logics, and *spatially* by activating domestic technologies to guide the use of the built environment, the people and objects within it, and the people and objects outside of it. As spatialized technology for home life, domestic futurity's primary function in the BR Smart Home is to architect a set of spatial relations and practices that bring domestic life in line with computational principles and emerging platform-based economic paradigms. Once online, domestic futurity helps to craft and subsequently normalize the ideal smart experience.

The spatial and technological narratives of the BR Smart Home help to distinguish the difference between old and new, inside and outside, stranger and self. In the BR Smart Home, rhetorics of technological innovation flex to fit the physical and figural formation of digitality in traditional and transitional spaces. The BR Smart Home promises to optimize domestic space by aligning the architecture and spatial rhetorics of the home to be coterminous with Amazon as a platform. Traditional rooms (e.g., the kitchen or butler's pantry) are outfitted with technologies that generalize the platform's capaciousness while also maintaining bourgeois domestic narratives about ease and luxury. These traditional spaces articulate digitality as a net positive innovation for domestic spaces, linking the next era of home with the normative ideology of the traditional domestic ideal. Transitional spaces (e.g., the drone landing pad) support the effort to shore up domesticity within the broader array of digitality, consolidating both technologies and pro-technology discourses into meticulously managed touchpoints with the outside world. The Smart Home augments the space's perceived functionality toward persistent computation as a way to maintain or optimize customary household functions.

However, the technological innovations orienting domestic futurity turn on nostalgic, conservative narratives of the home space as a private space wherein the external environment is subject to management, surveillance, and control.[48] The walls of the Smart Home and its attendant platformed features reify the divisions between gendered, racially, and economically distinct classes. Homeowners are discursively and spatially apart from stranger-others who are presumed threats, even if they are invited into the smart home by homeowners themselves. Such a domestic division, of course, is not novel or innovative. Rather, bias has been baked into the processes, functions, and expressions of domesticity since the initial iteration of the bourgeois domestic ideal.

In the BR Smart Home, digitality is the key to homeowners' technologically managed retreat—from labor, from worry, and from an uncontrollable (but ultimately necessary) domestic exteriority. Part of the luxury of this new domesticity is choice. Smart home owners decide when, how, and under what conditions to engage with the outside world. By design the choice is not everyone's to make; at the same time, when smart home owners elect to opt in to digitality, they also conscript a vast network of people, objects, and infrastructures who support their experience of living in digitality. The experience of living in digitality is available for purchase to the well-resourced digital subject, subject to the continued management of other digital subjects within the framework of luxury digitality. Whereas the homeowner is a favored and therefore protected digital subject worthy of luxury domestic futurity, the stranger-other is automatically suspect and subject to management by platform infrastructures. Domestic futurity necessarily implicates these stranger-others, who suffer surveillance and risk violence or injury to make the smart home's luxury experience possible.

The BR Smart Home leverages digitality as a sociocultural construct to manage users' expectations and broader cultural narratives about the types of domestic progress possible when living in digitality. Digitality quite literally shapes the architecture of these luxury-built homes while simultaneously architecting frameworks of value for computation within the contours of the domestic ideal. One of the household's normative functions is to calibrate its occupants' subjectivity and activity relative to others in society. The BR Smart Home supercharges the management of social relations through platform logics coded as luxury. Computational heuristics extend from the devices embedded in the walls to the broader domestic space. In this configuration of living in digitality, smart home technologies are pressed into service to a normative digital domestic ideal. BR Smart Homes employ smart technologies to define the utility of domestic spaces even as they articulate the social and economic relations between the household and the greater community. The infrastructure of domestic futurity is defined by two interrelated features: (1) corporatized platforms that manage the logistics of the BR Smart Home and (2) the contingent laborers who make smart domesticity possible. As a form of digitality, domestic futurity is defined by pleasure and protection for some and inherent risk for others. In this way corporatized domestic futurity reifies and reproduces a growing raced and gendered underclass materially separated from the more privileged social, political, and economic classes, that is, smart homeowners, those who can afford to outfit their home with technologies, and those who might pursue ease by plugging themselves into an increasingly global security state. For those outside of these communities of privilege, these leisure- and safety-producing devices may very well represent the opposite.

The global COVID-19 pandemic only exacerbated and intensified these divisions. As governments encouraged or mandated their citizens to stay home to stop the spread of COVID-19, the tremendous value of domestic labor—labor within the home and labor needed to make the home safe and functional—was exposed. Shortages in consumer goods, delays in shipping (including Amazon Prime two-day delivery), and rolling electricity blackouts evidence the importance of infrastructure, especially when vast swaths of the population are sheltering in place. At the same time, COVID-19 demonstrated the extent to which the home served as a locus for other infrastructures to converge. Maalsen and Dowling argue that "digital infrastructures have been critical to staying at home."[49] Although they focus their attention on digital technologies such as work-from-home software, contact tracing, and other apps, these scholars gesture to the ways in which domesticity is infrastructural in a broader sense: connecting homelife to the world beyond it. The rapidly proliferating smart home constitutes and condenses the very networks that afford opportunities to shelter in place, remaking domesticity within the corporate and digital framework of domestic futurity.

Of course, not all people could stay home during the pandemic. Stay-at-home policies, by necessity, implicate the need for accessible, safe housing and an entire class of workers to keep houses supplied throughout lockdown. During the pandemic, farmers, custodial workers, grocery store employees, delivery drivers, and more joined the ranks of healthcare workers as "frontline workers" who enabled others to stay home during the pandemic. These frontline workers took on great risk in doing so, and this risk was disproportionately borne by individuals living on the margins, with less access to social services such as healthcare. A report in *The Lancet* argues that "minority populations in the US disproportionately make up 'essential workers,'" suggesting that COVID-19 would exacerbate existing race and class hierarchies in the United States.[50] Amazon emerged as a major provider of goods and services during COVID-19.[51] Then, in October of 2020, Amazon reported that just under twenty thousand of its essential workers across Amazon and Whole Foods had contracted COVID-19 or were otherwise presumed positive.[52] During this time Jeff Bezos, Amazon's chief executive, increased his net worth by more than $90 billion.[53] Amazon also engaged in union-busting practices against workers organizing for better working conditions during COVID-19.[54]

Frontline workers were subject to additional surveillance, including thermal scanning temperature checks, to slow the spread of COVID-19.[55] The urgency and scale of COVID-19 created a state of exception that glossed over these facets of domestic futurity, rationalizing increased surveillance and risk to frontline workers as necessary for the good of the many. COVID-19 has "accelerated the integration and adoption of technologies within the home. . . . The same technologies that are enabling our lockdown lives now could normalize the heightened surveillance and control creep post-pandemic," which will be "experienced unevenly across housing tenure, gender, race, and class."[56] Now more than ever, we must interrogate the uneven effects of domestic futurity engendered by smart homes. Infrastructure offers a rich and nuanced framework to guide this work.

When domestic futurity is governed by the laws of capital and technology hegemons, smart homes are defined less by the cultural conditions of the community or familial relation that could define homelife. Instead, they are defined by the dominant features of living in digitality, namely, potential to secure increased profit margins, gained by automating or optimizing such relations. Smart homes also engage in surveillance and securitizing measures, which render almost all people outside the home subject to increased scrutiny, surveillance, and even violence. Still, surveillant technologies under digitality are inherently biased, and this bias is intensified by racialized histories of domesticity and housing. Contingent laborers and those stranger-others already suspect in the eyes of society disproportionately bear these risks.

The drone landing pad and the drop zone recalibrate domesticity by amplifying its hierarchical social and political relations using technological innovations within and beyond the home. At the same time, they spatialize both discursive and material technologies that mark off the interior from the exterior. This is why choosing to use smart technology is not an individual decision, but rather a collective one: securitizing strangers inside the home necessarily implicates those strangers outside of it. And although these technologies appear to be futuristic, their promises rely on reifying existing fear structures about the sanctity of the home. No matter their technological innovations, these smart homes cling strongly to signifiers of domesticity from the past, including gendered, raced, and classed visions of otherness. Traditional concepts of domesticity and domestic labor are ushered into the twenty-first century through platform capitalism and associated contingent labor practices, infrastructurally interconnecting "wider political economy . . . legal arrangements and standards . . . evolving cultural practices . . . and differential access to certain services."[57] Smart home platforms help constitute the technical and ideological build of the upper-middle-class smart homes that look and act like traditional suburban houses of a mythic American ideal. Like the American domestic ideal of eras past, only a certain class of people can access their pleasures.

As seen in this chapter, dominant conceptions of the smart home tend to prioritize efficiency and luxury for the nuclear family in normative configurations of domestic space. However, there are a number of smart home outliers that challenge these assumptions. The next chapter evaluates these outliers, showing how they modify dominant forms of domesticity typically promised by smart homes by experimenting with both spatial and cultural configurations of home. In so doing I show how these smart homes retell dominant narratives of the smart home norms and reimagine who and what is included in domestic life.

2

Smart Homes as Living Laboratories

I drove up to the Duke University Home Depot Smart Home (HDSH) on a gray, wet day in April. The dormitory was tucked into the verdant trees typical of this part of North Carolina. Even from within my vehicle, I could smell the scent of rainwater on pavement, rainwater on grass, rainwater glistening in the exhaling trees. Much of North Carolina is edenic, lush with green growth, sometimes for miles and miles on end. As a native Kansan who grew up in great expanses of prairie, upon moving to North Carolina as a doctoral student, I would drive through great green tunnels of trees with some feeling of claustrophobic trepidation. However, returning today, I was invited by the flora to breathe deeply and expansively on my way to the next smart home on my list.

Unlike the luxury, single-family Brookfield Residential Smart Homes encountered in chapter 1, this smart home houses a small group of university students who, in addition to completing degrees at Duke, live within an unusual and experimental smart domestic space. Their smart home experimentation has a focus: they are working to find, test, and document sustainable practices that could begin in the domestic sphere and echo outward—to other homes, to their immediate communities, and to even more public contexts. These students host periodic tours of their smart home dorm to publicize their work and invite collaboration with community partners. I joined them on one such tour on that April day in Durham. On tour I was prepared to hear about the technological features of the smart home. However, the smart home residents also invited me to consider my (and humanity's) position relative to the natural and built environment. Both the geographic location and the spatial rhetorics of the HDSH smart home encouraged such environmental introspection. This smart home's architecture and its residents make an argument for humans working intelligently, that is, symbiotically, with their surroundings.

The sylvan landscape surrounding the smart home certainly eased me into considering my role relative to the natural world. So, too, did the unusual architecture of the smart home. At first glance, the HDSH appeared to be less like a dormitory than a high-tech, industrial, postmodern commune. The two-story home boasted a slanted roof, pitched down to receive light in the large windows fitted under its apex. Like many who came before them, students in the smart home would endeavor to work with the natural environment; they would measure success by meeting the challenges of an unknown future partnering with surrounding communities and natural resources. Immediately viewable from the drive was a large water cistern, nearly a story tall, attached to the roof to collect rainwater. There were solar panels on the front porch next to wooden patio chairs. The front elevation was partially obstructed by vines climbing up a metal trellis. Fine yellow pollen settled on the porch and sidewalk had been disturbed by rainfall there. I parked among several vehicles lining the adjacent street and headed to the door.

This chapter considers two experimental smart homes, which, along with their residents and key stakeholders, demonstrate the need for just and inclusive domestic futures despite (and in response to) the political, social, and economic constraints of living in digitality. Whereas current configurations of living in digitality risk isolating families from their communities in a well-constructed technological cocoon, the smart homes we encounter in this book break down barriers between the public and the private spheres. Duke University's Home Depot Smart Home (HDSH) and Imagine! Colorado's Smart Homes (ICSH) are experimental smart home exemplars leveraging both new and existing technologies to solve public problems from the domestic sphere. HDSH is a live-in dormitory where students engage in sustainability practices to meet changing environmental needs. ICSH is a set of group homes that assist residents with intellectual and developmental disabilities (I/DD) in leading independent lives within a domestic community of care. These smart home exemplars characterize new possibilities for the smart home and provide a spatialized script to encourage the uptake of innovative domestic discourses.

Central to this revisioning of the smart home is an experimental and coalitional ethos augmenting domestic relations in space. As such, this chapter argues that these smart homes are *living laboratories* that mobilize smart technologies to construct new imaginaries for digitality in domestic and public spheres. Living laboratories are unique spaces where multiple stakeholders collaborate to solve problems specific to a community, place, or culture. For instance, living laboratories may work to encourage increased civic participation and deliberation in a bounded community, or they might map onto existing spaces to test and evaluate new ways to build inclusive local economies. The ICSH is a living laboratory operating at the intersections of advocacy, health care innovations, and accessibility efforts for disabled people at home and in public. The HDSH is a living laboratory that seeks to harness the capacities of young thinkers and campus community partners to live more lightly on the land. Each smart home designates the utility and purpose of its "smartness" through its particular affiliation with social issues and experimentation. One effect of this interconnection between smartness and purpose is that the aperture for what constitutes a smart home grows sizably, creating discursive and material space for new formations of smartness outside normative, corporatized, and exclusionary configurations of digital domesticity.

In this chapter I chart how smart home residents and their partners carve out new ways of living in digitality by experimenting with the space and functionality of the digital domestic sphere. The work proceeds as follows. In the first section, I introduce smart homes as a form of living laboratory. This section describes the features of living laboratories that stand in opposition to dominant conceptions of domesticity in general and smart domesticity in particular. In the second section, I perform a critical rhetorical reading of the living laboratories using the analytical frameworks of digitality introduced in earlier chapters of this book. This section offers a rhetorical analysis of the domestic spaces, drawing on three primary analytical inputs: the popular discourses that frame the ICSH and HDSH, the spatial rhetorics of the living labs, and the tours used to publicize their work. The analysis reveals that the living laboratory model of smart homes can help us to reframe the narrative around the utility and function of the smart home toward an ethic of care and inclusion.

Although these smart homes offer opportunities to refigure the form and function of domesticity, they also exist within a complex rhetorical situation that presents challenges to a complete reimagining. Class divides, unequal access to smart technologies or related resources, and broader social and political structures that manifest through scarcity narratives and exclusion can reproduce inequalities even in these exemplars. Therefore, each critical analysis of the living laboratory highlights rhetorical and material constraints to living labs' smart home innovation. Despite these limitations, these case studies show that champions of social change can use the smart domestic space to collaboratively solve systemic problems—particularly if the living lab model for innovation is expanded, centralized, and made more inclusive across the (smart) home horizon.

THE RHETORICS OF LIVING LABORATORY PUBLIC TOURS

As a rhetorician of space and technology, I understand public tours of smart homes as rhetorical condensation points in which arguments about the future of smart homes are communicated spatially as well as discursively. As explored in the introductory chapter of this book, public tours are a site for meaning-making and constructing a shared narrative about the subject(s) of the tours. Tourism can offer complex critique and found new political imaginations.[1] While they frequently communicate what has been, public tours can also communicate what will (or ought to) be.

As a rhetorical text, tours reveal the complicated relationship between public and private, the interior and exterior, and life at home versus life outside of it. Public tours of private smart home residences provide a glimpse into the interiority of the life held within while also making a public case for the mission, values, and features of the smart home. In the case of these smart home tours, tourism helps residents, key stakeholders, and tour participants make sense of emerging smart home technologies and their role in domestic and public life. In turn, tours can offer the discursive and physical space to imagine alternative, more just ways of living than the status quo.[2] At once, these tours (re)define the smart home, invite and persuade potential collaborators, and circulate messages about the role of technology in building the "good life." In the following sections, I evaluate key persuasive elements of public tours at two living laboratories, which suggest possible futures for inclusive and care-oriented smart home living. I augment this analysis with interpretive readings of related discourses, including public-facing promotional materials for the smart homes and related media coverage of the smart home living labs.

Living Laboratories

How we describe the smart home matters. Language framing the purpose and utility of the smart home sets up expectations for how a smart home—and its residents—ought to behave. Any discursive framework can also settle into the spatial arrangement of the home itself. As Shannon Mattern has argued, "metaphors give rise to technical models, which inform design processes, which in turn shape knowledges and politics, not to mention material" spaces.[3] For example, a computational metaphor structuring the smart home might reward efficiency efforts. Rooms become sites of ordered rationality; their domestic perfection is achieved with the fewest steps to the fridge, through a particularly ordered arrangement of objects that automate away life's excesses, or through the application of an administrative mind-set to everyday living.

But there are other ways of imagining the smart home. The homes featured in this chapter perform an alternate intelligence through the spatio-discursive enactment of their "living laboratory" designation. Before turning to the smart home living labs individually, let me first offer a framework for living laboratories more broadly. Living laboratories are *experimental spaces where multiple stakeholders collaborate to solve problems specific to a community, place, or culture.* They are distinct from other types of laboratories because they incorporate everyday activities into the experimentation. The literature on living laboratories treats them in one of three different, yet often overlapping, ways—as place, process, and method. Some scholars suggest that living laboratories are either a city/town, geographic region, institution, or building in which experimentation occurs. For instance, Carrol and Rosson have studied "community informatics" projects in "university towns," elsewhere known as "wild places," where these community-based, experimental processes unfold.[4] So, too, Veeckman and van Der Graff envision smart cities as living laboratories.[5] Living laboratories gesture to the unfixed and processual nature of crafting new place-based spatial imaginaries. As such, another modality of the living laboratory foregrounds the *processes* in which stakeholders are empowered to experiment through, among other creative modalities, "iterative . . . cocreation processes."[6] "Living laboratory", then, may not only be a noun (in the sense of place) but a verb (in the sense of experimenting.) Finally, living laboratories are at once spaces and methodologies for community engagement and citizen participation.[7] Scholars interested in researching with or for specific communities may use living laboratories as methodological scaffolding for discrete problem-solving specific to a population or a public.

As a concept and set of spaces, the living laboratory is overdetermined, and its true essence (if there is such a thing) is likely somewhere among these three definitions. My own definition of the smart home living laboratory highlights the inherent liminality of the living lab as a social and technical space that *imagines* and *enacts* a novel form of living in digitality. Whereas hegemonic forms of living in digitality foreground capitalist accumulation, efficiency, and profit over all else, living laboratories' iterative processing invites communities to figure spaces toward other ends. When understood as a collective, community-held, and justice-led experimental process toward public innovation, living in digitality takes on a different patina as a process of caring for self, others, communities, and environments. While this chapter treats smart homes as living laboratories, the concept of a living laboratory is not specific to the smart home era. Depending on which scholar you ask, living laboratories are either a few decades or nearly a century old.[8] Either way, living labs have long served as spaces where targeted, in-place experimentation incorporates diverse stakeholders to benefit wider communities.

Whereas other smart home logics condense power and knowledge in the hands of a few users who pay for the privilege, living laboratories seek to share their spatialized knowledge practices in new spaces. A key component of the living laboratory is to mobilize smart technologies to provide solutions to problems of collective import. Once solutions are tested in the smart home spaces, they are made available to wider publics. Modes of dissemination are multiple, and tours are one important medium for transmitting learned knowledge. Smart home tours reflect on the invitational processes of living laboratories. They work to persuade outside stakeholders of their role in building domestic futures anew. The dissemination of smart solutions and the hailing of possible collaborators are central to the living laboratory process.

Living laboratories are particularly worthy of rhetorical study in the field, as they are place-based processes that are concerned with understanding and communicating the meaningful, liminal nature of everyday living. Analyzing these living laboratories from an "embedded" rhetorical perspective highlights the material processes of worldmaking hosted in smart homes designated as such.[9] Actively refiguring living in digitality occurs within the smart home spaces themselves through discursive and spatial rhetorical processes explored throughout the rest of the chapter.

ICSH and HDSH

The ICSH and HDSH are smart homes constituted in spatial and public discourses as living laboratories, characterized by localized, justice-centered, issue-specific technological experimentation within a domestic space. Their spatial narrative is organized in part around the experimentation inherent to the laboratory setting. By incorporating this experimental function toward inclusive public problem solving, living laboratories augment the operationalization of digitality. Whereas corporate technology hegemons currently frame the value of smartness through capitalistic efficiency or increased consumption, living laboratories offer a new heuristic for intelligence. Here, smart is not an exclusive (or exclusionary) cognitive capacity located naturally in some bodies though not others. Nor is smartness figured as a particularized identity trait that necessarily positions certain bodies or spaces over others. This ableist proposition assumes in advance a binary figuring of intelligence and ability embedded in a broader ontology of deficit and lack.[10] Rather, intelligence is framed in its inherent multiplicity. It is a recognition—and celebration—of the many intelligences of individuals, communities, and the spaces they create and encounter together. And in the case of the smart homes we encounter in this chapter, smartness is defined inclusively as capacity for care—for self, for others, for the public, and for the planet.

As sites of discursive experimentation and persuasion, living laboratories help create new meaning-making structures for both domestic spaces and the emerging technology that mediates the smart homes' existence. Living laboratories offer a set of spatio-rhetorical tools to advance arguments about the smart home space as a site for social change. By localizing social problems, living laboratories facilitate the formation of new narratives and spatialized practices for smart homes. Living labs are, in turn, important sites of material change in the social world, because they subsequently move discursive and spatial experimentation on key concerns "beyond . . . laboratories and design studios, and into the world."[11] As other smart homes purport to do, the ICSH and HDSH center the needs, desires, and ambitions of their residents. Unlike other smart homes, however, they simultaneously encourage the knitting together of discrete domestic spaces with community efforts working together in kind. The ICSH and HDSH serve populations beyond their particular domestic interior by encouraging collaboration with community partners, transmitting results from live-in experiments to the greater public. The ICSH and HDSH perform the living laboratories designation in the material spaces they operate in even as they discursively reconfigure "home" in the context of higher education and housing for disabled individuals. In so doing they challenge the dominant definition of the smart home as a site primarily for luxury and efficiency.

The smart homes' enactment of the living laboratory paradigm frequently determines its "smartness." Public discourses linking "smart" with the living laboratory comprise various news media, press releases, public statements and interviews, and the websites of the ICSH and HDSH. For instance, the ICSH argues that its "SmartHomes are serving as *living laboratories* to design and test technologies that may then be moved out to individual or family homes, so that individuals with intellectual and developmental disabilities may remain in more independent and natural settings versus more costly and segregated settings."[12] So, too, the HDSH is constituted as a smart home precisely through its designation as a living laboratory. Larry Montea, vice president of student affairs at Duke University from 2001 to 2019 , argued that the HDSH "is a *living laboratory*" in which "students actually reside . . . and live with the benefits and consequences of their technology design and deployment decisions."[13] In this way the home is "a testing ground for student research and product innovation," with impacts that are felt beyond the HDSH.[14] Importantly for both the ICSH and HDSH, the smart home is configured as a living incubator for site-specific problem solving, but it is not the end of experimentation. The HDSH and ICSH contend that solutions found in their discrete spaces may be incorporated in larger contexts to benefit the greater community. Public tours are one modality for sharing learned insights necessary to rework living in digitality to benefit the community.

In the following section, I perform a site-specific rhetorical analysis of the ICSH and HDSH using information gathered during the tour and public discourses about the living laboratories. I show how these smart homes enact the living laboratory mission by innovating domestic solutions to public problems in private spaces. However, these smart homes do not operate within a vacuum. Rather, they are subject to internal and external forces that challenge their capacity to innovate, including a shrinking social safety net. The following paragraphs reveal smart homes responding to resource scarcity in real time. Such scarcity limits the living labs' transformative potential and makes it more likely that living labs exist in sites of relative privilege. In the following sections, I also highlight how these smart homes manifest experimental advocacy positions in complex, sometimes combative, political conjunctures. Thus, although these living laboratory smart homes seek to serve comparatively different communities, they share a need to be innovative within political and economic constraints that could limit the reach of the smart home innovation.

THE DUKE UNIVERSITY HOME DEPOT SMART HOME

A small, round sign installed in front of the rainwater collection tank informed me that Smart Home tours began in the front of the house. I joined a dozen others making their way through the front door. Bunched umbrellas were upturned outside the door, under the roof's overhang, next to angled black solar panels. Expansive windows and glass-fronted doors allowed visitors to peek inside the covered porch. Once inside, we were received in an open area anchored by a long wooden table and a galley-style kitchen. Shuffling and slightly damp, I introduced myself to one of the student guides greeting tour guests. I surmised that the several tour guides were all students at Duke University. They would shuttle my fellow tourists and me through this living laboratory, past clearly visible and ongoing scientific experiments, and art exploring humans' connections with the natural world and the built environment.

Like other laboratories, smart home living labs often specify a problem, function, or mission to direct their work. As Veeckman and van der Graff argue, "living labs can have different thematic focuses and interests, such as focusing on innovation, in health, media, smart grids, etc."[15] The Duke University Home Depot Smart Home's thematic focus is smart sustainability efforts. The residents and collaborative partners of HDSH use the smart home to engineer solutions to climate change, mitigate its deleterious effects, and work harmoniously with the natural environment through everyday lived experience. For example, the HDSH has built landscaping and gardens meant to steward natural resources. Smart sensors embedded throughout the landscaping help students or facility workers evaluate the characteristics of soil, and they trigger the irrigation system.

The HDSH's mission is a prescient, urgent one, given that climate change is one of the most pressing existential crises facing humanity. Presently, humans are ill equipped to respond to the urgency and intensity of these challenges. As Sander van der Leeuw and colleagues note, "we face many urgent sustainability challenges, including climate change, loss of biodiversity, poverty, epidemics, and violent conflicts . . . [that] are complex beyond our understanding, outside the capacity of our current institutional structures to address, and caused by ingrained behavioral patterns."[16] Responding to sustainability challenges therefore requires urgent, innovative, transformational practices not currently in practice. Any solution must incorporate diverse communities and stakeholders, it must be participatory and active, and it must not only focus on diagnosing problems but "become a venue for collaborative and rigorous exploration, testing, and implementation of solution options."[17] The HDSH promotes a form of this "sustainability science" by leveraging the strengths of academic research and community partnerships in an applied, embedded context.

The HDSH is unique compared to other academic institutions *and* living laboratories, in that it is a smart dormitory, part of Housing and Residence Life at Duke University. Student residents drive the functioning of the house, working with external partners and faculty advisors from engineering and other programs. The HDSH houses approximately ten students in six thousand square feet, with a majority of space dedicated to shared spaces.[18] Shared spaces include areas one might expect in a domestic space, including a kitchen and dining room. More atypical domestic spaces include multiple laboratories set up for different types of experiments. Whether lab space or living space, shared space is at a premium, because the Smart Home is also the site for collaborative projects with the university, including the interdisciplinary, university-wide Smart Home Club. Although residents and Smart Home Club participants engage in sustainable practices beyond the home, the HDSH is the epicenter for prolonged, applied, in situ experimentation that can echo out to relevant communities. Ultimately, the HDSH meets the designation of living laboratory insofar as it is a domestic residence housed in a preeminent research institution that invites collaboration from community members and corporate partners.

Living in the HDSH is competitive, limited to less than a dozen students at any one time. Students must apply to join the ranks of the smart home residential program and pay residential fees consistent with on-campus housing. In 2018–2019, the annual fee to live in the smart home was approximately $9,000 per person. In addition to financial contributions to the university, the students must demonstrate their value to the smart home mission. Once accepted, they are expected to not only live in the dormitory but "contribute as a member of the Executive Board or of a project team, and as a tour guide."[19] While it is true that students who live at the HDSH are likely to experience institutional and social privilege by virtue of their social and physical location at Duke University, they also provide added value in terms of their labor and their willingness to open up their lives to curious outsiders who wish to witness the smart home in action—on regularly scheduled tours.

I am introduced to these students as tour guides and residents on one such tour. My student tour guide notes that they no longer reside in the home but are still involved in the leadership and administrative function of the house. Early in the smart home living laboratory tour, my student tour guide begins to articulate the smart home's definition within the living laboratory mission. As my guide notes, the HDSH emerged from a student's vision to have an on-campus, live-in laboratory smart home.[20] It was decided that this domestic experiment ought to have a mission or function beyond simply being "smart." The smart home should be considered smart because it has utility beyond flashy technologies such as home automation. Rather, the smart home should be evaluated by its ability to support sustainability within a broader community effort.

The tour guides and the spaces make arguments for what the smart home is and could be. The tour and tour guide suggest that the smart home mobilizes existing and emerging technology to respond to climate change concerns. On the tour, two key narratives are offered about the future of the smart home. The first narrative foregrounds collaboration with others and with the environment. The second suggests that transparent engagement with others is necessary to meet sustainability needs. Both narratives are operationalized through the content of the tour and reaffirmed in the spatial rhetorics of the HDSH. In so doing the HDSH redefines what "smart" means. Smart technology is not the endpoint for this living laboratory; rather, it is a tool to mobilize students, their partners, and the broader community into a relation of care and concern about each other and the environment.

Narrative 1: Smart Means Sustainable Science through Collaboration

As the tour proceeds, I understand that the home communicates the value of smart, sustainable partnerships both discursively and spatially. The collaborative narratives baked into the smart home tour are reinforced through spatial discourses of the smart home and the everyday lived expression of the smart home's interior. For instance, the home's unique shape is communicative of its residents' values and supporters. The strategically placed windows and roofline, I learn, regulate temperature naturally. Solar panels are distributed around the space. The massive rainwater collection tank is the first thing one sees upon visiting the house. The panels and the tank are aesthetic and functional objects, signifying that the home uses natural resources through smart conservation. The domestic but industrial aesthetic of the space also gestures toward aesthetic minimalism, corporate partnerships, and the farms that operate just miles from this space. The spatial rhetorics of this smart home suggest that architecture can be in service of a higher good. Here, as in other domestic spaces, the architecture supports the life inside of it. Moreover, the HDSH lends itself to supporting the life outside the home—the natural world working in harmony with its human-built counterpart.

Figure 4. Duke University Smart Home. Courtesy of the author.

Much of the front elevation is composed of floor-to-ceiling windows outlined in white against the grey siding. As my tour guide notes, the house is perfectly oriented according to the cardinal directions. The situatedness of the building allows the home to capture sunlight efficiently, providing natural light for the house. These windows are also key to the temperature management of the home. Letting in the sun in the winter helps regulate temperature, warming the space naturally. However, in the hot, humid Durham summer, the sunlight can overwarm the home and its residents. The smart home's residents have developed a low-tech but highly effective natural strategy here. Partially blocking the windows on the front of the smart home is a series of trellises. In the summertime, grapevines crawl up these trellises, blocking the sunlight and cooling the interior ambiently. In the winter, the grapevines die away, leaving more sunlight to shine through. When I visit in April, the vines are just rousing from their slumber. Soon, they will grow tall, stretching their limbs to the top of the support structures provided them.

The spatio-rhetorical message is this: the built environment, the flora, and the climate ought to work together symbiotically in theory and in practice. This message is communicated by the material architecture of the home, which connects the interior to the exterior. To be sure, the windows, trellises, and vines provide functional but natural management of the light and, subsequently, heat. When taken together, they also function rhetorically as an argument for sustainability. The vines materially signal an intention to work with the natural environment rather than against it. Using natural methods to regulate temperature and light invites the natural into proximity with smart home residents. The smart home space is shared; the mission to protect nature is reflected in nature's protection of those inside of it. On the rainy day I tour the home, I sense the interconnection of humans and the natural environment as the soft, filtered light invites me into the space.

The interior design mirrors, rather than opposes, the external environment. Most of the year, the large windows usher in the lush greenery enjoyed by Durham, North Carolina. The HDSH is nestled among a number of green trees, which are clearly visible from the interior of the smart home. The bountiful organic elements are equally reflected inside. Cork flooring is under my feet, a more sustainable option than other wood-based flooring options. Wooden elements define much of the first floor. The large common room where guests are received features a large wooden table, capable of seating perhaps ten to fifteen people. A galley-style kitchen is outfitted with light, wooden cabinetry. Where there are no cabinets, the walls are painted a light green color that reflects the poetry of trees outside. The glass wall across from the cabinets makes one feel as if one could reach through it to touch the plants. I am decidedly inside a space protected from the wind and the rain. Nevertheless, this domestic interiority is only *just* demarcated from the exterior space. Residents and visitors are positioned as partners inextricably linked with an environment that blends the goodness of the external world seamlessly into the domestic space. Together, natural elements pair inside and outside to support the sustainability mission. They also communicate the mission of the living laboratory: Here, nature is nurtured. In turn, it nurtures those inside the HDSH. This space tells a story in which humans living in digitality coexist in harmony with their natural environment. Here, technology's definition is purposefully expansive. For instance, solar panels are deployed to decrease energy use. Rainwater cisterns, which date back several thousand years, are hardly an emerging technology.[21] Yet they are categorized as "smart" insofar as they contribute to the sustainability mission and other intelligent climate control systems. By giving technology a higher purpose, "smart" begins to move beyond hegemonic, technoliberalist rhetorics about automation or corporate profit. Instead, smartness becomes designated as a meaningful descriptor for collaborative, mission-based science.

Cooperation and partnership are continued themes of both the smart home space and our tour of it. Students using technology are positioned by the tour as active participants in creating a more sustainable future. However, as my tour guide notes, they cannot do it alone, which structurally encourages experimentation and the development of collaboration within and beyond the home. Thus, the second definition of smart articulated on the tour gestures toward *smart partnerships within the local community*. Here, the home is lent its *intelligence* from the students in it, the prestige of the school, and the smart partnerships with community stakeholders. External partners, including major brand organizations, offer a form of institutional legitimacy that lends their resources and reputations to the home. By organizing the space around student experimentation and partnerships with stakeholders, the smart home modifies "smart" to include advancement of human knowledge. This advancement narrative is, upon first glance, both politically neutral and rhetorically unambiguous: when "smartness" has a focus, especially an altruistic one, the smart home becomes a lever for positive change.

Even as residents and visitors are invited to join a partnership with the natural environment, the HDSH's location on a university campus gestures to the partnership between university stakeholders and smart home residents and visitors. In this way the tour connects many audiences through the rhetorical construction of shared values, including cooperation, symbiosis, and partnership. Spatially, the smart home constructs an audience out of disparate strangers who may share an interest in technology, sustainability, or Duke University. Here, the smartest expression of the smart home gestures to the interconnection of humans, institutions, and the sustainability mission. This smartness is organized discursively through the tours and spatially through the smart home itself. In the following paragraphs, I show how the smart home refracts the register of "smart" through material and discursive transparency, signified by wide-open architecture and glass.

Narrative 2: "Smart" Means Shared: Transparent Partnerships by Design and Necessity

As I tour the home, I am struck by how light and bright this smart home is, even on an overcast day. The lofty, open feeling is granted by tall windows on the home's front elevation. Nearly a full wall of windows encases one whole side of the shared living and dining area. Above me, light from large glass windows tucked under the eaves streams through the open loft area of the second floor. I am experiencing the exterior world through the smart home's interior. Even so, I am in plain view of any passersby.

Smart home living laboratories are different from other labs insofar as they serve both public and private functions in the domestic sphere. They manifest communication and collaboration between groups of people in a localized space. As Nesti argues, "the concept of open innovation" is central to living laboratories. Open innovation is characterized by "a continuous process of exchange of knowledge between actors and of learning-by doing."[22] To enact this mission, living laboratories demand literal and figurative transparency. Therefore, transparency must serve a second but related function: to invite others to actively participate in the experimental processes of the living laboratory. Almirall and Wareham note, "living labs operate by involving users in explorative processes in real-life environments."[23] Outside community members must be able to participate in the public problem solving that occurs in the otherwise private spaces of the living laboratory. What's more, the public must perceive the living laboratory as invitational as well as transparent. They must see themselves in the living laboratory's mission and the spaces used to enact that mission.

The HDSH's built environment obliges. Transparency is quite literally figured into the design of the home. The open-air architecture of the HDSH built environment communicates this value in a markedly spatial way as it invites external shareholders inside. Figurative and spatial transparency renders smart home residents subject to "community partnerships" with any member of the community nearby. The windows similarly perform an important figurative function of the experimentation typical of a living lab in this space. These floor-to-ceiling windows show the everyday lived experiments that occur in the space. The transparency and structural significance of the many windows that make up the first floor invite stakeholders into the space literately and figuratively. The HDSH supports these explorative processes by serving not only as a site for domestic experimentation but also as a site to witness those experimentations from within and without.

Spatial configurations of openness gesture to the possibility for residents living in digitality to innovate within and alongside human and nonhuman others inside and outside the smart home. Insofar as the scientific process turns on transparency as a value and methodology, living laboratories must be open and transparent to communities beyond them. As Almirall and Wareham note, living labs are "collaborative projects engaging companies, academics, government and technological centers, where users are involved in nascent development stages and successful iterations are validated in real life environments."[24] The Duke University Home Depot Smart Home invites community partners to join in the experimentation process and support the sustainability mission through the open, accessible built environment. Spatially and infrastructurally, wider publics are invited to witness the processes unfold inside this smart home and even to become active participants in it. The HDSH turns on collaborative processes and shared experimentation with external stakeholders like other living laboratories. In turn, the HDSH as a living lab makes a covenant with the public: to serve them in the public sphere and the private.

Engaging multiple stakeholders is central to the functioning of a living laboratory, which by definition encourages community and corporate partnerships. At the beginning of the tour, when I explained that I was a researcher from Kansas State University, I learned that my public tour had fallen during an alumni weekend at Duke University. In fact, I later determined this tour was held during homecoming. These tourists had likely learned about the HDSH from a list of prescheduled homecoming activities. My tour, then, was unique in two key ways: first, it is likely that a larger number of tour participants were present than is typical on such tours. The increased number of tourists necessitated several guides leading several tours at once. Indeed, the built environment reflected this condition: the space was full and frenzied, pushed beyond its usual daily paces. Second, the audience for the tour was composed mostly of alumni returning to Duke as part of homecoming activities. My savvy tour guides appropriately appraised this rhetorical audience; the alumni were ultimately framed on tour as potential partners in innovation. At many higher-education institutions, alumni become important agents in supporting academic processes.[25] I gathered this tour was one way to capture their imaginations and entice partnerships with—and perhaps financial or other types of support for—the university.

While smart home tours serve as a dissemination point for learned knowledge, they are also a persuasive and invitational method for engaging potential partners beyond the university. Corporate sponsorships are one significant example of this practice. As noted in the tour, the original smart home resulted from a $2 million investment, primarily from Home Depot (thus the name.) Other sponsors, however, included General Electric, Siemens, John Deere, and more. Like other living laboratories, corporate sponsorship aims to create a mutually beneficial partnership between the university and its corporate partners. A list of early sponsors is prominently displayed on the home's first floor. The particulars of the arrangement were not described in great detail on tour, but the subtext was apparent: students gain access to experimental space and equipment, and the corporate partners receive tax benefits and access to students from one of the nation's most prestigious universities.[26]

That the Duke University Smart Home is elsewhere known as the Home Depot Smart Home evidences interest in (or the necessity of) developing partnerships with external stakeholders, including alumni or private interest actors. The HDSH is not alone is such contexts. Not-for-profit organizations frequently work around the generosity of others, receiving goods and services donated to them and making them work as best as possible. However, the structure of the funding model means that there may be a disconnect between what is needed and what is available. In the case of the Duke University Smart Home, one of the larger donated items was their heating, ventilation, and air conditioning (HVAC) unit, which resides in the basement of the HDSH. The HVAC unit is an expensive and important piece of equipment in the modern home because it controls ventilation and temperature. HVAC units also play a major role in the energy efficiency of a house; modern, energy-efficient systems use less energy but are costly to buy and install. HVAC units must also be appropriately sized for a home. If they are too big, HVAC units "will use more energy and will not remove humidity adequately," which may "lead to mold and mildew problems" and "causes a lot of wear and tear" on the HVAC unit.[27] Finally, an improperly sized HVAC cannot effectively perform its heating and cooling duties, leading to significant temperature oscillations and hot and cool zones in the house. The lack of temperature consistency can invite human intervention into the system, wherein residents fiddle with the thermostat in a futile attempt to control the temperature. Taken together, these problems make even an energy-efficient HVAC unit energy-inefficient.

My student tour guide notes that the HDSH was gifted a "very big [HVAC] system" that does not properly fit the space. Although this gift reduced overall initial construction costs, the improperly sized HVAC unit ultimately made the HDSH less energy efficient than if it used a properly fitting one. In a smart home tasked with reducing energy usage as part of a community sustainability effort, energy efficiency is crucial. As my tour group descends to the basement of the HDSH, my fellow tourists begin asking questions about energy efficiency. Is the house energy neutral, or does it produce energy for the grid? How much power do the photovoltaic and other solar panels produce? My tour guide notes that at times the HDSH is less energy efficient than other homes. As of 2016 the smart home could generate only 10 percent of its energy needs.[28] The overly large HVAC is "constantly fighting against itself" to "reach an equilibrium that it can never reach" in a system for which it isn't fitted. Moreover, the solar panels used in and around the house are too old to be efficient. As my student tour guide notes, the HDSH has not yet found a corporate or other partners to support the purchase and installation of new solar panels. Taken together, these two conditions mean that the sustainable home is sometimes less energy efficient than yours or mine might be, undercutting its claims toward sustainability.

Economic constraints certainly undermine the transformational capacities of the HDSH as a living laboratory. Most generally, smart home residents are devising and carrying out experiments with the wrong tools. But there are broader concerns with making the proper functioning of a living laboratory contingent upon a corporate partner—or set of corporate partners—for funding. As a private institution, Duke University's funding models draw on multiple sources, including student tuition, endowments, and other funding and resources from external stakeholders, such as donors, alumni, and corporate partners. Even at the wealthy and exclusive Duke University, with a self-reported endowment of $12.7 billion and nestled in progressive Durham, North Carolina, the tour suggests that funding challenges remain for valuable projects like the HDSH. To some extent, this is to be expected: higher education's increasingly corporate financial framework revolves around the assumption of a supposed (often constructed) resource scarcity and a subsequent need to "trim the fat" of academic programs and projects that don't meet a certain set of ever-increasing (but often black-boxed) bureaucratic metrics. Such a narrative mirrors digitality's incessant desire to root out "inefficiencies" through the assumed "rational" logics of capital. Once standardized, the deficit model of university funding structurally discourages public investment in the university and instead encourages financial privatization at all levels.

One effect of this particular framework of value is that champions of the sustainable smart home are tasked with perpetually pitching the project on public tours to public and private interests that may or may not share their vision for enacting sustainability. As a living lab, the HDSH and its residents are equivalent to a petri dish under a microscope for potential investors. The architecture of the smart home space inherently figures public ways of living for residents, who are understood to be ambassadors for the living lab mission. To be sure, this covenant asks residents to give up some of their privacy in exchange for a unique, service-based living situation. Although those on tour are kept primarily to "public" spaces—we do not see the students' dorm rooms, for example—it is clear that these students are expected to perform an ambassadorial role for the smart home and for the university. This is a special type of labor, one that asks students to trade some of their social capital for the benefit of others (and the environment).

It matters here that young people are the ambassadors of the smart home mission. Throughout the tour we are guided not by administrators or professors associated with the project but by young adults with an established relationship to the smart home. The symbolic suturing of youth culture to smart technology and, relatedly, its promise is a powerful rhetorical move. Scholars have shown that young people have nuanced relationships with technology, despite oversimplistic cultural expectations about their presumed connection to it.[29] Still, youth culture is a powerful reservoir for meaning and social capital, not only because young people metonymically stand in for the future. Associating a particular project with youth culture provides a sense of (post)teenage cool that can be assumed to transfer to the project itself. By foregrounding the "next generation" of scholars, activists, and domestic innovators, even overtly technoliberalist projects can be imbued with both gravitas and a marketable futuristic orientation.

Even when partners are found who support the overall mission of the smart home, living laboratories outline a particular *process* for achieving sustainable goals that may not align with corporate interests. Relatedly, smart home champions must repeatedly justify the existence of a program whose end goal is not immediately capturable by rationalist metrics endemic to the corporate world. In so doing smart home champions must translate the logics of environmentalism, care, and collaboration inherent in the sustainable smart home ideal into a quantifiable language consistent with the logics of accumulation or profit. This leads to some awkward rhetorical questions: How, precisely, can one quantify the value of a healthy home or healthy planet? How does one quantify (or, the next step, *monetize*) inclusive community partnerships for a healthy and resilient environment? Ought it matter whether there is a business case for climate justice?

Another effect of this funding model is that it encourages corporate greenwashing of university spaces, including the smart home. Greenwashing "enables companies to present themselves as caring environmental stewards, even as they were engaging in environmentally unsustainable practices."[30] Big-box stores like Home Depot, the HDSH's namesake, which in the past claimed to be "the world's largest buyer of construction material," has marketed itself as a green retailer.[31] As a retailer, its eco-friendly solutions seem in large part to be based on maintaining or increasing consumption—not making or buying fewer products but making or buying "greener" ones. Other sponsors of the sustainable smart home, like Siemens, have been accused of supporting "a mega [coal] mining project that has breached environmental protection rules."[32] Engagement with a promising project like the HDSH might provide rhetorical cover for companies with less-than-savory environmental histories, to the detriment of the larger goal of climate justice. The transparency baked into the discursive and spatial architectures of the living laboratory model may be perverted into a tool affirming capitalistic practices, where both budget models and everyday domestic practices are oriented around logics of profit or, relatedly, scarcity. In this way, the smart home both resists and is stitched into larger technoliberalist frameworks of value, space, and domesticity. This leads us to an important, if awkward, question: If even the most-prestigious and well-resourced universities are unable (or unwilling) to support social experiments for good absent corporate sponsorship, how else are less-resourced or underserved groups supposed to do so?

The mismatch between donated resources and the critical mission of the home is not the only limiting factor with which the HDSH and its student advocates have to contend. Rather, the HDSH's sustainability mission sometimes directly contradicts local, regional, and even national goals and governance. While Duke University is part of a politically progressive part of North Carolina (sometimes referred to as the Research Triangle, so named for the many local research universities and institutions), the state is frequently conservative and even regressive as it pertains to climate change. North Carolina's legislature has explicitly rejected climate change science and, by extension, climate change as a scientifically knowable truth. In 2020 experts suggested that Carolina's coasts were likely to be influenced by climate change, particularly sea-level rises. These scientists produced a report that suggested that thousands of miles of Carolina coast were at risk from over three feet of possible sea-level rise. The risk is intense and imminent; the Carolina coast was already changing and could be underwater before 2100. In response, according to local newspaper the *News and Observer*, legislators "essentially banned the use of the 100-year sea-level-rise forecast by many of the agencies the state controls."[33] One potential justification for the ban was that the report would influence the coast's seasonal tourism and property values. North Carolina boasts both mountainous and coastal regions. Duke University is almost equidistant between them. Tourists enjoy these attractions in great numbers. In so doing they supported the state's economy to the tune of over $25 billion in 2018.[34] Coastal communities are a sizable part of North Carolina's tourist industry.[35] Climate change threatens the coasts, including sea-level rise and major hurricanes that can disrupt the lucrative tourist industry in increasingly unpredictable ways.[36]

North Carolina will likely serve as a harbinger of climate change effects for the rest of the United States. The HDSH seems ideally positioned to innovate climate change solutions in a potentially unstable geographic region. However, the HDSH's advocates must do so carefully, as climate science is not accepted universally across the state or, indeed, the nation.[37] The unfriendly political and social climate is particularly troubling because of the funding structure of climate-related nonprofits. Given a shrinking social safety net, nonprofits may have a tenuous time convincing external stakeholders that they are worthy of support. While the HDSH offers useful domestic solutions to climate change, there are also risks to straddling the public-private world. In the case of the HDSH, the risks of publicly disseminated and externally sponsored domestic experimentation are considerable. The HDSH risks alienating itself and its university from local publics and the state legislature in pursuit of science to counteract humanmade climate change, a charge which some dispute. This alienation may lead to less resource support to sustain its vital mission. In a world where higher education increasingly relies on external funding support, the HDSH must strategically position itself as valuable and worthy of investiture, in a risky political climate.

Taken together these conditions produce a unique rhetorical situation, as the very future of the smart home depends on persuading external stakeholders of its utility. On one hand the HDSH offers an aperture for developing transparent, collaborative, community-minded solutions for the impending climate crisis. The smart home spatially and discursively reconfigures "smart" as mission based, with a focus on care and sustainable practice. The living laboratory model is potentially innovative and transformative, as it supports experimental practice and care for self and others as a portable principle for a new form of living in digitality. Partnerships with external stakeholders can mean that solutions found within a living laboratory may be distributed beyond the living laboratory. From this perspective, "knowledge is diffused within society" such that "new solutions to problems can come and should be collected from inside to outside an organization."[38] However, the sustainability mission of the HDSH is limited by economic, logistical, and political constraints, including constructed narratives about financial scarcity that seem to necessitate corporate funding for progressive projects. Ultimately, the HDSH must find innovative ways to champion sustainability science in response to and, sometimes, defying these constraining factors. The smart home tours are at once a way to argue for the mission of the HDSH and invite potential collaborators to sustainably participate in its mission. The next section shows how another not-for-profit smart home innovates within similar constraints.

IMAGINE! COLORADO SMART HOMES: THE CHARLES FAMILY SMART HOME

I visited the Charles Family Smart Home (CFSH) in Longmont, Colorado, in the middle of June 2019. The day was grey with pockets of light rain as I drove from my hotel to the Denver suburb of Longmont. The town is about thirty miles away from Denver and twenty miles away from Boulder. The CFSH is tucked away in a tidy, well-kept suburban neighborhood. The neighborhood is a recently developed one, bespeckled by single-family bungalows and modern Craftsman-style homes. The CFSH's front elevation is similar to its neighbors. However, it is distinguished by its size, which appears to be twice that of its closest neighbors. Its roof is outfitted with solar panels, and natural wood shingles cover much of the house's face, gesturing to the environmentally friendly scene characteristic of this area of Colorado.

At the front door, Leona Stoecker greets me warmly. An "emeritus" member of the Imagine! Colorado Foundation board, and former mayor of Longmont, Colorado, Stoecker plays the role of Imagine! evangelist. As we wait for all tour participants, Stoecker gives me and the rest of the small tour group a rundown of the basic features of the home. The CFSH cost, I learn, was $1.5 million. Its development was made possible by several sponsors (as in the HDSH, their names are

Figure 5. The Charles Family Smart Home. Courtesy of the author.

on the wall nearby), including a $100,000 kickoff contribution from the Charles family, for whom the house is named. The home has recently celebrated its tenth anniversary. At the beginning of the tour, Stoecker rejoices that this and the other smart home mortgage is paid off, in part through her fundraising efforts. "I'm always asking for money," she notes. The economics of CFSH will remain a central narrative element on this tour, as the smart home caretakers labor to maintain a high quality of life for residents in a challenging socioeconomic landscape. Imagine! Colorado's will to serve the community was abundantly clear. Nevertheless, the budget realities are always at the forefront of the mind—and the tour.

ICSH are Colorado-based group homes that leverage technological advancements, community partnerships, and domestic experimentation to support their inhabitants. Their residents include individuals with intellectual and developmental disabilities (sometimes referred to as I/DD or IDD) and, on occasion, those who care for them. In Longmont, Colorado, the Charles Family Smart Home (CFSH) is one of two smart homes supported by Imagine! Colorado residential services. It is open to tours by the wider public. The other smart home, also named for its benefactor, is the Boulder-based Bob and Judy Charles SmartHome. Both smart homes were developed and are presently managed and administered by Imagine! Colorado, a nonprofit whose mission centers on "creating a world of opportunity for all abilities."

Imagine! Colorado is a designated private, nonprofit community-centered board (CCB) founded in 1963 by parents of I/DD children. According to the Colorado Department of Health Care Policy and Financing, CCBs "support access to long term services and supports (LTSS) through Medicaid waivers for

Home and Community Based Services (HCBS). CCBs currently provide case management, entry point functions, and HCBS for individuals with intellectual and developmental disabilities."[39] In other words, Imagine! Colorado connects those with I/DD and their families to support and services in the region.

In addition to serving in this role, Imagine! Colorado offers several direct services for I/DD individuals. These services are "designed to incorporate people with developmental, cognitive and physical challenges into the fabric of their communities."[40] For instance, Imagine! Colorado serves individuals through all stages of life with employment services, adult day programs, school-age services, mental wellness services, and services for infants and toddlers.[41] In 2018 Imagine! served over four thousand Coloradans and their families through various programming.[42]

Especially pertinent to this chapter are Imagine! Colorado's Residential Services, which serve approximately 165 I/DD individuals in the greater Denver/Boulder metro area. One way in which Imagine! manifests its mission to serve is to develop and maintain innovative residential services and care spaces for individuals with intellectual and developmental disabilities. These residential services support several living modalities, including in-home family models, host home models, and group homes. The Charles Family Smart Home in Longmont, is an example of the group home model, which supports residents with many diagnoses and needs. The Charles Family Smart Home presently serves six live-in residents, and the Boulder Smart Home is the home of eight I/DD individuals.

Both ICSH are experimental sites whose technology-based solutions may be disseminated to the wider community. Like the HDSH, the ICSH serves as a living laboratory to revision the meaning of both "smart" and "home." According to the Community Foundation of Boulder County, the ICSH are among the first smart homes "in the nation to use technology to increase the self sufficiency of people with intellectual and developmental disabilities and to augment the efficacy of caregivers."[43] Like the HDSH, the ICSH facilities routinely partner with local community organizations and institutions, including the University of Colorado, Boulder (CU). At CU, for instance, first-year open-option students work with ICSH residents to engineer domestic solutions to support residents' well-being and agency. In this way, the smart home performs the living laboratory designation by inviting external stakeholders, including students, to participate in their domestic experimentation. Students in turn receive training. Sometimes they take this training into careers in technology accessibility and design postgraduation, building out more inclusive futures.

Spaces and spatial design serve as monuments to social attitudes toward disability and are mechanisms that proscribe exclusionary social attitudes in space. The ICSH offers one model for inclusive domestic design, using smart technology nimbly to support the needs of disabled residents. By intentionally serving underserved communities and "hacking" existing smart home products to support disabled individuals in community-based environments, the ICSH, its residents, and its champions advance alternate narratives for the smart home. The ICSH constitutes a new form of smart domesticity by reframing intelligence as inclusive and agile.

Narrative 1: "Smart" Means Inclusive: Domesticity for All Bodies

Housing and domestic design figures disability *spatially* through the material configurations of the home and *communicatively* through discourses surrounding the domestic ideal. Generally, there is insufficient housing available to support individuals with disabilities, including those with varying physical and cognitive needs.[44] Disabled people are traditionally left out of conversations about housing equality and design. Even critical analyses of domesticity and housing, which consider race, class, and gender, tend to exclude disabled populations as relevant stakeholders. As urban studies and disability scholar Rob Imrie notes in the editorial of a special issue of *Housing Studies* on disability, domesticity, and housing, "disabled people and their needs have rarely been featured in academic or policy debates about the nature of housing quality and conceptions of housing quality have rarely identified disabled people as of significance. . . . the subject matter is seen as insignificant to the understanding of social patterns and processes; that is, disabled people are part of a minority group. This view is not only found in parts of housing studies, but reflects a broader, societal-wide, indifference to the subject matter of disability."[45] For Imrie and others, disability is a socially constructed phenomenon in which impairment is figured pejoratively, as lack, often in relation to space.[46] From a sociospatial perspective, disability is not necessarily a characteristic specific to an individual and their abilities and needs; rather, disability is the "product of social and environmental processes which constitute people as disabled."[47] Many of these processes are spatialized and perpetually reified in the design of public and private spaces. Housing itself may be "inherently disabling or enabling . . . through its physical characteristics, administration and finance."[48] In other words, the built environment can inscribe exclusionary social attitudes into spaces. As Elizabeth Guffey puts it, "through graphic communications, urban planning, and architecture, design weaves social attitudes into the very shape and texture of our built environment."[49] In other words, architecture and environmental design renders exclusion and inaccessibility material in space.

In response, disability scholars, activists, advocates, and policy makers are evaluating the myriad ways in which housing and the related concept of domesticity reflect and amplify exclusionary attitudes in space. Some disability scholars suggest a wholesale reimagining of domesticity from the perspective of disability. Such a project is a natural extension for advocates of critical disability studies, whose proponents suggest that "disability is *the* space from which to think through a host of political, theoretical, and practical issues that are relevant to all."[50] In a recent edited collection on the relationship between disability and domesticity, disability historian Michael Rembis argues in favor of "disabling domesticity." This critical project demands the rejection of disability (and domesticity) as a monolith, calling instead for a "complexly embodied understanding of domesticity firmly grounded in its material realities and the dominant socioeconomic structures that pervade disabled people's lives."[51] Disabling domesticity "creates a politicized theoretical space from which a disabling critique of dominant heteronormative ableist constructions and material manifestations of domestic life can be articulated."[52] From this perspective, the household and its attendant configurations of domesticity are inflected with social and political meaning about the nature of the body, ability, and space. Disabling domesticity recognizes that homes are always already sites of political action and, for disabled people, not always (or even often) private.[53] Moreover, because housing is a liminal space between public and private, domestic space becomes an important site of encounter for apparently "external" forces, including housing policies, social attitudes, medical practices, socioeconomic structures, and more.

The modern notion of the American dream home discursively and spatially figures both domesticity and disability through a white, economically privileged, able-bodied, and heteronormative lens.[54] Disabling domesticity means troubling cultural orientations to both domesticity and disability to create new theoretical vantages and new material configurations of space that are accessible and inclusive as a way to reimagine "home, care, and family."[55] It means qualifying the domestic ideal as a product of an ableist society. Here, the domestic ideal is a social construction that requires ongoing discursive maintenance even as it manifests in very concrete ways for disabled bodies. Key to this work is recognizing the ways in which space can (re)constitute hostile conditions for disabled individuals in public and in private.

As a project, disabling smart domesticity necessitates the recognition of the inherent paradox of an automated home. On the one hand, smart technologies can be a useful tool for disabled folks to communicate, monitor their health and wellness, connect with friends and families, and more. On the other hand, the notion of smartness as an idealized framework for living draws on and reasserts ableist logics of embodied value regarding "intelligence." As we saw in the first few chapters of this text, the dominant definition of smartness in the smart home aligns "smart" with efficiency or increased productivity.

As a capitalist metric, smartness architects intelligence and ability in inherently exclusionary, individualistic, and moralistic ways. In a 2021 analysis of "smartness" rhetoric in public health campaigns, Emily Brooks notes that the "language of intelligence" calls upon eugenicist histories in which medical professionals would use IQ tests to diagnose and categorize disabled people. Such categorization was not simply medical; intelligence was articulated within a framework of morality wherein "immorality and stupidity are conflated."[56] Perhaps the most glaring example of narrative ableism in the smart home is that the so-called alternative to the smart home is often referred to as the "dumb" or "stupid" home. Accessibility advocate Jennifer Kretchmer has similarly argued that "the word 'dumb' is an ableist slur rooted in bigotry. . . . It equates being nonverbal with a lack of intelligence and furthers stigma around the disabled community."[57] Because language architects reality, ableist language matters in the production of smart home space.

Viewing smart domesticity through the lens of critical disability studies is crucial insofar as these vectors influence the built environment of the home in ways that include or exclude some bodies at the expense of able-bodied individuals.[58] Because smart domesticity is rendered intelligible in narratives about smartness as well as in space, disabling smart domesticity means recognizing how smart homes reaffirm exclusionary domestic ideals in discourse as well as in the built environment. For instance, Andrew Marcum explores the "dialectical relationship between an ongoing movement for disability rights and the built environment."[59] (Re)configuring the material conception of the home toward an accessible, inclusive politics can provide grounds for individuals and communities to gain access to needed social services, care, and kinship.

In the case of living laboratories, domestic experimentation is often the first step to the dissemination of inclusive solutions. According to Imagine!, the ICSH act as "living laboratories to design and test technologies that may then be moved out to individual or family homes, so that individuals with intellectual and developmental disabilities may remain in more independent and natural settings versus more costly and segregated settings."[60] Imagine! Colorado smart homes are a form of spatialized worldmaking as they work to transport their vision for accessibility and a high level of care beyond individual smart homes.[61] Imagine! envisions these smart homes as experimental spaces in which the domestic sphere might be reinvented to meet the needs of more and different types of residents, no matter their abilities.

Narrative 2: "Smart" Means Agile and Savvy: Hacking Low-Tech or No-Tech Solutions

After all members of the small tour group arrive, Stoecker introduces us to the technology lead, Christopher S. Baumgart, or, as Stoecker calls him, the "genius with the computer stuff." Baumgart, who asks us to call him Chris, shies away from the term. Monikers like that make him "very nervous," because it makes his job "sound like magic." His job, he notes, is to show the "pros and cons" of the smart home. Nevertheless, as we proceed through the tour, I learn that if Baumgart has a particular genius, it is not necessarily based on high-tech skills, like the ability to patch together a mesh network. Rather, it is the ability to hack low-cost technology and low- or no-tech modalities to support residents with varying needs and diagnoses. Much of the tour clarifies that this skill is necessary given the political and economic conjuncture in which CFSH operates.

Chris takes over as the lead tour guide in the first few minutes of the tour. We proceed into a spacious living area, where a few residents are engaged in activities. Before Chris introduces one of the residents, he asks for permission to share information about them. This resident, we learn, "does not communicate verbally." Rather, they communicate through assistive technologies, including dedicated, technological communicative devices. At this moment, however, Chris asks instead for the resident to give consent to the tour through hand signals. This is an opportunity to learn about the various modalities for serving I/DD individuals with various diagnoses. But more importantly, it's a framing mechanism for the rest of the tour. The message is this: the home is built around the needs of its residents, featuring a variety of mediated and technological communication channels and solutions. Not all of them are digital. This low-tech (or no-tech) theme repeats throughout the tour. It quickly becomes clear that the smart home is smart not because of its technological capacity but because of how it centers its residents with an ethic of care. Here, smart technology does not exist for its own benefit but is valued because of its dexterity in supporting its residents and caregivers.

Nevertheless, we are on a smart home tour and presumably have an interest in the technological aspects of the CFSH. Thus, Chris begins to discuss the assistive technology devices' function and the resources required to obtain them. As Chris notes, each assistive technological device costs around $6,000. The high price presents accessibility issues for many families; even though medical programs like Medicaid will cover the cost, the approval process for coverage is long: a "six-months process," says Chris, who adds that this extended timeline entices the manufacturers of the device to "amp up the price" to make it worth their while. If and when a resident is approved for the assistive device, they must use it for five years before Medicaid approves another device.

Mandating such a prolonged use presents a significant challenge for residents, my tour guide notes. First, assistive devices are not guaranteed to last five years, even with the best of care. The manufacturers will warranty them for only three years, leaving a two-year gap in coverage. Second, rapid changes in technology can render even the smartest devices obsolete within a few years. Chris gives the example of the iPhone, which, as he notes, has been around for only thirteen years and has produced a dozen models, each with significant technological updates. If an iPhone breaks, Chris notes, the smartphone can be repaired nearly anywhere. However, if a Medicaid-approved assistive communication device breaks and is not under warranty, a resident or caretaker must begin the long and arduous process of Medicaid approval again.

Chris suggests an alternative smart technology at this point in the tour. Where pricey Medicaid-approved devices fail or break, off-the-shelf consumer electronics offer relatively low-cost success for residents. The iPad, for instance, can be a stand-in device for other dedicated communicative technologies. Software and hardware packages can perform nearly every task the dedicated assistive technology can, at a much lower price. Perhaps even more importantly, it can be fixed by nearly anyone certified to repair Apple devices. For this and other reasons, off-the-shelf technology is an attractive solution for Imagine! Smart Homes, their residents, and their caregivers.

Although he does not use these terms explicitly, Chris is gesturing toward "hacking" both the technologies and the politico-economic systems that provide structures of support (or not) for disabled individuals. On the one hand, a truly inclusive social and medical system would offer disabled people not only useful communication technologies but agency in selecting and using those technologies. A truly inclusive built environment could make some standalone technologies obsolete or at least less requisite to engage with an existing space. On the other hand, ICSH and their champions are tasked with providing the best possible living experience for their residents given the existing climate. In the interim, then, lower-cost and more widely available technologies serve as a fairly effective stop-gap solution while still maintaining a high level of care for residents. The ICSH's successful use of off-the-shelf technology around the home evidences the temporary feasibility of this solution and provides an entry point for advocacy. As we learn on the tour, Imagine! has successfully lobbied for Medicaid to cover iPad devices for its residents and others who might need them. In this way, Imagine! and the ICSH, in particular, are seeking creative technological solutions in response to political and economic roadblocks.

Another example of this hacking ethos is presented later in the tour. This particular home has been fitted with a call-and-response system throughout the house, including the bathrooms. One function of the system is to help caregivers log interactions and support services. Another function is to give residents the agency to use the bathroom in private and call for support as necessary. What Chris calls the "price tag" for such a critical system is astronomical—$60,000 each; installed across both homes, $120,000 in 2019 dollars. Furthermore, like other technologies, this one is aging and getting buggy. As Chris notes, the call-and-response system was becoming resource intensive rather than efficient; their IT support person needed to fix the system almost every day. The technology made to lighten the load was instead contributing to it. Funding to update the system is difficult to come by, and even if it were achievable, it would likely take a long time to get approved through bureaucratic channels.

In response Chris had retrofitted the home with off-the-shelf Amazon Echos, which served as command center and hub. These Echos are paired with several devices throughout the house. Some residents used special color lightbulbs to remind them to perform daily activities. Others used them to automate daily routines. In the CFSH, Echos pair with low-cost Amazon Echo buttons, which look a bit like game show buzzers. Residents use them in the bathrooms to signal when they need assistance, effectively replacing the aging technologies hardwired into the walls. In this way the CFSH has effectively responded to economic scarcity and political gridlock by working around the long Medicaid approval process, costly medical equipment, and half-life of technology. The CFSH becomes a platform for advocating for smart and flexible medical technology policies, with some proven success at making procedural changes in gridlocked political contexts. In the meantime, the CFSH residents retain access to the services that support dignity, self-sufficiency, and agency.

The tour reveals that this performance of the good life scaffolds onto several aspects of the smart home living laboratory. First and foremost, the CFSH performs the main designation of the living laboratory: experimentation in space. Residents and caregivers engage in intricate, studied, and imaginative domestic experimentation using technological and cultural constraints as a starting point. The scientific process is enmeshed in IT solutions that rework ready-made technologies to address persistent problems through highly regulated, hyper-specific assistive technologies. Residents cannot wait to live their lives until policy makers make commonsense laws about assistive technologies. Hacking off-the-shelf smart devices is a stopgap solution, but it is a feasible one for the CFSH. Moreover, these consumer-based solutions are eminently transferable to broader publics. I/DD individuals may benefit from these hacks in the present tense. Living labs, characterized by open innovation, can transmit learned knowledge within and beyond the domestic sphere. These experimentations can serve as the basis for directed political advocacy. The CFSH performs all of these functions, through partnerships with outside stakeholders.

The CFSH reorganizes the register of "smart" to include diverse populations. It offers access to "the good life" for individuals with varying diagnoses by building more-accessible domestic spaces. To be sure, these smart homes deploy technologies that may presently be characterized as "smart." The CFSH has modified smart technology to serve residents both generally and in highly specific, need-based contexts. However, what makes the CFSH smart home truly intelligent is not automating lights. Rather, the smart home is smart because it reconfigures domesticity to benefit people of all abilities. From this vantage point, the ICSH's suite of smart homes is an exemplar not just because of its technological capacity but because of the way the spaces reimagine homelife for some disabled residents and their partners, families, and communities. Most generally, Imagine! Colorado's residential facilities stretch the modern conception of home to be more inclusive, adaptive, and savvy, creating processes that respect their residents' dignity, independence, and privacy. Imagine! works with residents, caregivers, policy makers, private citizens, donors, and others to provide and advocate for an independent and accessible homelife for individuals of all abilities. The ICSH in Longmont and Boulder manifest this mission to serve spatially in the format of the living laboratory.

Technological advancements in the home can be life-changing for many different types of people, no matter their disability. However, the introduction of domestic technologies does not automatically result in a techno-utopia. Most generally, access to useful technologies is not equally distributed in a highly segmented society. For disabled folks, and especially those who are nonwhite, not economically privileged, or alienated from communities of support others enjoy with relative ease, smart technologies can be out of reach—to varying degrees. Scholars have argued that disability is "deeply connected to technology" with both technical and "radically relational dimensions."[62] Feminist technoscience scholar Laura Mauldin calls this interconnection between societal privilege and technologies of care a sociotechnical "entanglement."[63] For Mauldin, although "care work being performed by caregivers is increasingly medically complex, ongoing, and intertwined with and using knowledge and tools associated with science, technology, and medicine," not all have access to these technoscientific advancements within current spheres of social, technical, and medical relation.[64] The domestic (and public) experience of disabled individuals is continually reconfigured by this differential access, such that technologies aimed to support disabled folks have a negative effect if they are not adequately distributed.

The Charles Family Smart Home is an exciting but complex site for disabling domesticity. Attending to disability from a sociotechnical perspective requires a deeply situational perspective. Health and wellness are among the most popular use-cases for smart technologies, but the able-bodied individual remains centered in design of both smart technologies and smart homes. What is needed are smart home exemplars who provide instructional roadmaps in centering disabled residents, their needs, and their desires as standard practice of domestic care. The ICSH facilities operate at this unique intersection of technology, disability, and care, through the enactment of the living laboratory paradigm, though these technologies and care networks are not universally accessible. Disabling domesticity from a smart homes perspective therefore mandates a consideration of the interlocking political, social, and economic structures that figure both disability and domesticity.

THINKING SMART HOMES ANEW: LIVING LABORATORY PARADIGMS

Presently, smart homes can represent little more than branding-in-space. The smart home of the status quo is largely defined by corporate actors and technology hegemons who produce smart technologies. At best these smart home living laboratories rhetorically create possibilities for reimagining the home as a more egalitarian, inclusive, and experimental space. They define—through action and great care—what smart homes *mean*. Living labs enable space-based, experiential experimenting. They enact transformational politics by inviting members of the polity to engage in practices of placemaking aligned with the topical focus of the experimentation. When organized in community partnership, smart home living laboratories may offer literal and figurative blueprints for reorganizing the modern domestic. Here, digitality's distributed and optimizing frameworks rest in the hands of community members, who, when they recognize the biased effects of technology, can countermand the technoliberal frameworks of value that construct the modern smart home.

In this chapter I analyzed the spatial rhetorics of two smart homes, the Duke University Home Depot Smart Home and the Imagine! Colorado Smart Home, as presented through two public tours. I advance the argument that the HDSH and CFSH recalibrate smart domesticity by not only serving their residents but architecting scalable solutions for climate change and exclusionary health care practices. Both tours not only suggested a novel and inclusive definition of smart homes but rhetorically positioned the *intelligence* of the home as a discursive and spatial domestic configuration of care, innovation, and inclusivity. Here, the intelligence of a domestic space is defined through its ability to augment social, political, and economic realities in the living spaces. In so doing, those identified with the experimental smart home define "smart" homes differently—not as corporate commodities controlled by technology hegemons who develop smart devices but as spaces to mount a challenge to unjust structures around sustainability and inclusivity.

The ICSH and HDSH are cause for optimism about our digital domestic future. These smart home living laboratories spatially and discursively reframe the hegemonic forces of modern digitality, calling into question the definition of the smart home as primarily an exclusive, efficient, and consumptive domestic structure. Whereas dominant conceptions of "the smart home [are] ideologically underpinned by normative homeownership and the nuclear family whose ontological orientation is mediated by the digital," the smart homes featured in this chapter strategically challenge key elements of dominant smart home discourse.[65] Smart home living laboratories are key sites of technological and cultural liminality insofar as they utilize new technologies to "disturb the incumbent activity system, raising new possibilities and challenges."[66] Spatially and discursively, the smart home living laboratories constitute sites of experimentation to unstitch domesticity from the nuclear ideal.

The spatial rhetorics of these experimental smart homes can help to reconfigure the structure of relation standardized by dominant modes of digitality and, by encouraging radical reimagining of the home, mount a challenge to exclusionary narratives of (smart) domesticity. One mechanism for inciting social change from the private sphere is to challenge the definition of public and private, capitalizing on domestic technology's capacity to unsettle through experimentation. As sites of experimentation and creation, the smart homes and their residents invite introspection about the function of domesticity and digitality and ask probing questions about who ought to belong inside the modern smart home. In turn, these residents and their homes rework the smart home space to reimagine social relations and foment public engagement within the domestic sphere. By altering both the footprint and function of a smart home, living laboratories create conditions of possibility to challenge technoliberal ideals of living in digitality that continue to organize the present and future definition of the smart home.[67]

Many of the living lab's spatio-rhetorical experimentations are successful. The HDSH has been especially adept at imagining and documenting symbiotic relationships between humans and their natural environment, drawing on the affordances of architecture and public engagement to encourage sustainability. The CFSH succeeds in remaking off-the-shelf commercial technologies for the benefit of its residents, challenging technical and medical structures that assume, in advance, an able-bodied user. Both have seen success operating as a site for advocacy and change making around sustainability and disability. Both are exemplars for creating sites for public engagement.

But they do so within a complex and, at times, combative sociopolitical environment. Living laboratories' domestic innovations remain constrained by digitality's structural capacities to shape the material and discursive world of smart homes (and of capital more broadly). Both the rhetorical appeals used by the living labs' champions and the spatial arrangements of the smart homes demonstrate the long, pernicious reach of technoliberalism and platform capitalism. As a generalizable and expansive cultural condition, living in digitality is inherently *adaptive*, rerouting even significant alterations in the smart domestic space to comply with its broader aims of optimization, control, and incorporation. The ICSH and HDSH offer important lessons for challenging the smart home status quo. What is needed now is further circulation of these ideas, increased support for their missions, and resources to support additional innovations across inclusive and diverse communities.

While the living laboratory provides a technical, spatial, and discursive template to reimagine domesticity anew, the designation and practice of the smart home living lab is not a panacea. As we have seen in this chapter, the radical potential of smart home living labs is constrained by several forces specific to living in digitality in the present tense. Hegemonic expressions of living in digitality are, by their very nature, biased toward spatial arrangements that prioritize consumption, expansion, alienation, and extraction for the direct or indirect benefit of platform capitalists. Moreover, the smart home living laboratories and their residents are embedded within biased cultural and economic systems that restrict efforts to reimagine the discursive and spatial arrangements of the smart home.

As such, the radical transformations sought by smart home stakeholders are necessarily limited in range and scope. As the case studies in this chapter show, these smart home living laboratories require a significant number of resources to support their critical missions. And as the case studies show, access to necessary resources is not guaranteed.[68] Like other not-for-profit endeavors, these experimental smart homes are affected by a socially constructed climate of resource scarcity whose discursive architecture encourages competition rather than collaboration and affiliation. Frequently they rely on private or commercial solutions to counterbalance an increasingly small social safety net. For instance, the Duke University living lab's capacity to reduce energy usage is constrained by donated appliances that are not adequately fit to the home's specifications. The Charles Family Smart Home is a community project, but is indebted to the largesse of a handful of major donors. The rapid rate of technological advancement, particularly in the smart home, frequently means that residents, caretakers, and other stakeholders must consistently seek external support to keep the smart homes functional and online.

Even when resources are available, they are not widely distributed across all communities. For instance, one of the living laboratories evaluated here occupies a privileged position at a well-resourced, prestigious, private research university in the American South. Even then, the exceptional opportunity to live and work in a smart home incubator is limited to a handful of students at an already exclusive institution. The other living lab is benefited by a small but tenacious community rejecting normative standards of able-bodiedness in home and technology design, while agitating toward equality for disabled individuals from the domestic sphere—a relatively progressive social geography enables that agitation to take hold in discrete locations. To be even more effective, there should be more smart homes like them, and they should be adequately funded to carry out their missions across socioeconomic, geographic, and institutional contexts without having to rely on corporate partnerships. A truly transformational effort to refigure living in digitality toward more just ends will transport the most-inclusive and culturally intelligent features from these living labs to the mainstream smart landscape, a topic explored in chapter 4 of this text.

The next chapter continues to investigate the meaningful nature of smart homes. Instead of focusing on discrete smart home spaces, I instead look at some of the technologies that support the flexible smart home. In these configurations smart homes are created by incorporating off-the-shelf smart home products embedded with artificially intelligent virtual assistants Siri and Alexa.

3

The New Old Housewives

In January of 2020, the New York Times *published an article advising its readers to whip their homes into shape. The article advocated a specific domestic technology: virtual assistants Siri and Alexa. Titled "Put Alexa and Siri to Work," the piece was structured as a step-by-step guide to employing the many artificially intelligent virtual assistants (AI VA) on the market today. The goal: running an efficient and tidy home with the wonders of smart, personable technology.*

In "Put Alexa and Siri to Work," virtual assistants are positioned as laborers—an extra set of (virtual) hands to make domestic work easier. Assistants, the article notes, "can make life more efficient" because "they can help you ease into the day and keep the household running smoothly." Virtual assistants are labeled "helpers," with "powers" to run the household, control gadgets, aid in shopping, and even feed the family. Yet, in this article, Siri and Alexa are characterized as more than just disembodied robot workers. Assistants are given another, more abstract role—as "digital companions" with which one could share a homelife. From this point of view, virtual assistants are useful in the home not only because they can help the household run but also because they are a helper, friend, and intimate. In an alienating and uncertain world, Siri and Alexa are situated as willing and trustworthy domestic partners. This NYT article, then, is an "introduction to life with a digital companion," or, in our terms, how to live in digitality with the help of Siri or Alexa.[1]

"PUT SIRI AND Alexa to Work" is not only a set of instructions about using a technological tool. Rather, it is an instruction manual on living in digitality by incorporating an increasingly humanized AI into everyday life at home. Personifying technical tools with humanlike characteristics is nothing new. Nor is it particularly novel that popular media offers householders advice on how to run the home given changes in society or technology.[2] However, in the current moment, popular discourse about virtual assistants has moved from instructing about incorporating devices to prescribing the *relationship* users should have with them. This situating of humanlike machines is part of a collective rhetorical process to clarify the meaning of the virtual assistant and its role in the social and domestic spheres. Virtual assistants like Siri and Alexa are written into the ongoing narrative about living in digitality using discourse consistent with the larger symbolic economy of digital domesticity. By relying on humanistic tropes to describe virtual assistants, domestic advice media increasingly encourage the *normalization* of AI VA personification. Moreover, the humanization of virtual assistants calls into question their authority and agency as rhetors in the digital space.

Explainers like the *New York Times* article are a discursive response to changing material and social conditions in the household as well as cultural shifts in technological adoption. They reflect a world in which hundreds of thousands of Americans are likely to be living with a virtual assistant. Many already do. AI VA are increasingly pervasive in the modern American smart (or becoming smart) home. A 2017 study from the Pew Research Center suggests that "nearly half of U.S. adults" use voice-controlled digital assistants.[3] Given consistent growth in the adoption of smart home technologies ahead of and during the global COVID-19 pandemic, that number has likely only grown since 2017.[4]

One of the biggest drivers of digital assistant usage is the proliferation of smart devices and other connected products with AI VA built in. Virtual assistants are available in myriad devices, from phones to smart speakers, laptops, and microwave ovens. These devices are branded not only by the companies that produce them but also by their particular embedded AI VA. Moreover, they are associated with a specific technological ecosystem. Nearly all of the major players in the smart home space offer their own virtual assistant or are compatible with those that already exist. For instance, the Apple suite of software and hardware provides access to Siri; Alexa services Amazon's smart home suite; and Microsoft features assistant Cortana. Because they are included in smart devices and are associated with existing technological systems and products, they are part of a vast platformed infrastructure with significant logistical power. As a result, "Siri, Google Assistant, Alexa, and Cortana have quickly become household names," partly because they have moved into many American households.[5]

Virtual assistants are an important entry point to the smart home experience because they promise ease, convenience, and control of the domestic space—while also creating a hospitable and companionable environment for users. For instance, promotional material from Apple suggests that "Siri lets you stay connected without lifting a finger."[6] In website copy for its virtual assistant Alexa, Amazon argues that "Alexa makes your life easier, more meaningful, and more fun by helping you voice control your world."[7] From a user perspective, these voice assistants can be useful for executing household tasks. They may also be perceived as "cool" or trendy in addition to functional.[8]

Yet there is another reason to explain virtual assistants' rise in popularity and use, and it has everything to do with how AI VA are positioned in the smart home. In addition to executing basic tasks, AI VA are programmed to have a charming, disarming, and familiar personality.[9] Smart devices with AI VA built in are smart and capable and also friendly and companionable, "managing both data and emotions, that is, both the logistical and affective negotiations within a home."[10] AI VA will perform household tasks efficiently and with a good sense of humor. The narrative characterizing the service of AI VA is amplified in promotional media and popular discourse about the assistants. In addition to running one's household, AI VA will—in the words of the *New York Times* explainer—become a user's "digital companion." In this discourse virtual assistants are as functionally flexible as they are efficient; they are "caring, nurturing, responsive, attentive, helpful but not willful, smart but not overly so, replaceable, customizable, available."[11] By all accounts, virtual assistants are willing and able to serve their users in domestic space across contexts, whether their needs are logistical or more emotional. As I have argued throughout this text, living in digitality is an immersive computational condition categorized by the spatialization of digital technologies and the expansion of bias and control in technical format. AI VA connect the world of the technological to the social to make living in digitality approachable for the masses. AI VA are, therefore, a carefully constructed on-ramp to living in digitality.

The on-ramp is well used. Siri and Alexa have become so familiar to the American public that they have become familial intimates in the modern domestic imaginary. Writing for the Brookings Institution, for instance, Mishaela Robison observes that virtual assistants "practically become members of the family—conveying information, telling jokes, and controlling home appliances."[12] To be clear, they are not just any member of the family. By virtue of the domestic labor they perform, their patterns of speech, and their location within the smart home, we can more accurately say they are the modern-day, techno-optimized *housewife*. Like mythic housewives of the domestic past, these artificially intelligent virtual assistants will perform domestic labor, including care labor, with a (virtual) smile.

In this chapter I figure Siri, Alexa, Google Home, Cortana, and the cadre of other virtual assistants as "the new old housewives." New old housewives perform gendered, raced, and classed labor at home, materializing regressive stereotypes of domestic workers through novel technological frameworks materialized in space. As new old housewives, AI VA help transport the biased ideological perspectives of digitality to the home space through the performance of their raced, classed, and feminized work and their carefully configured affects that (re)calcify existing spatial narratives about women and domestic laborers. Smart homes are constructed in the cultural imaginary as futuristic, even as dominant versions of digital technologies affirm the traditional domestic ideal. Virtual assistants manage the contradiction between technological futurism and traditional domesticity by performing traditionally feminized tasks, including household chores and care labor. Hitching virtual assistants to the durable but flexible figure of the housewife explains the persistence of stereotypical tropes in AI VA, despite voice adjustments or changes in the device's name. In turn, the modern smart home's usage of AI VA spatially reorganizes smart domestic labor through platformed efficiency, intensifying the reach of platform technology in intimate spaces. Highlighting AI VA's gender trouble is important because gendered AI VA can "devalue women's traditional roles in society and exacerbate violence toward women."[13] In other words, how we treat AI is likely to influence how we treat people, including women, nonbinary folks, femmes, and domestic workers more generally. My argument here is not only that the language that we use to describe AI VA matters, although that is true. Even more than that, the language used to articulate the purpose and utility of AI VA is a vestige of a gendered heteronormativity that underlines domesticity writ large. Smart homes spatialize this idealistic domesticity and subsequently create a home for it in the public imaginary now and in the future. How we figure AI VA evidences that digitality is agential, constraining future language, thought, and action around digital devices and performances of gender at home.

The following case studies trace the entanglements among discourse, labor, femininity, domesticity, and space. Guided by the theoretical apparatus of *living in digitality* outlined in the introduction of this book, I use a spatiorhetorical lens to read a collection of artifacts, smart spaces, and multiauthored discourses about Siri and Alexa. These artifacts include technology reviews and reports; how-to guides for engaging with the smart home; advertisements for smart home products; smart home applications and skills; and more. I also examine how Siri's and Alexa's narrative framework and discursive capacities are located in the smart home space through individual devices, networks, and spatial configurations. This constellation of discourses tells a complex but increasingly clear story about the gendering of AI VA, one which mobilizes traditional discourses about home while at the same time pitching futuristic narratives about how to live in digitality with the same old gender politics.

Part of my broader argument regarding virtual assistants in this chapter is that these devices offer an inviting entry point into a more general smart home system—and digitality. Said another way, *living in digitality* as an immersive condition is made possible by the rhetorical architecture of AI VA as feminized domestic agents. The anthropomorphized aspects of Siri and Alexa are key features of the modern smart home in particular, and living in digitality more broadly. The increased use of AI VA and their strategic positioning in the cultural imaginary also suggests increasing mechanization of the domestic space as computational and therefore subject to optimization and management.[14] In our language of digitality, Siri and Alexa generalize the experience of computation by relocating it in a friendly, domestic figure with features that appear to mirror a particular mythologized domestic subject—the racialized, middle-to-upper-class, suburban housewife. Of course, the new old housewife is indeed a *myth*. More than that, she is an *impossibility*, a time-traveling and idealized figure whose possibility is betrayed by a nostalgic yearning for a prior temporal and spatial order. She is also an impossibility because her privileged status is possible only given uneven social architectures that manifest through the laboring bodies of domestic workers, paid and unpaid. Still, as a category of meaning, she is useful for evaluating modern technologies because virtual assistants are—and continue to be—made in her image.

This chapter proceeds in three sections. In the first I briefly draw together relevant literature in three primary fields of study: the history of domestic technologies, analyses of feminized labor at home and work, and critical-cultural analyses of AI VA. The second section explains how virtual assistants have proliferated against a backdrop of smart technologies, using gendered (and therefore inherently raced and classed) stereotypes as a social lubricant toward increased smart home adoption. In the third section, I examine recent attempts at disarticulating gender from virtual assistants. I conclude by explaining why the spatial rhetorics of smart homes make the prospect of genderless virtual assistants especially difficult, at least in the context of the modern smart home.

THE NEW OLD HOUSEWIFE

In this chapter I use the term "new old housewife" to explain how the gendered persona of AI VA fits into a historically specific narrative of housewifery, a set of actions, behaviors, and performances that transport the smart home of today to the twentieth, nineteenth, and even eighteenth centuries, and back again. More generally, the figure of the *wife* is a useful heuristic to document, categorize, and problematize the gendered, raced, and classed narratives embedded within discourses *about* virtual assistants and virtual assistants' own rhetorical performances in domestic space. Scholars use the term *wife* to explain the capaciousness of AI VA's gender problems and their myriad discursive and material manifestations. Kylie Jarrett, for instance, examines the intersections of domestic labor and digital media capitalism in the figure of the "digital housewife," whose feminized work is often invisible, undervalued, and underpaid.[15] Yolande Strengers and Jenny Kennedy have recently written a book on the rise of the smart wife.[16] This chapter adds to the growing collection of scholarship linking domestic technologies with the sociocultural formation of the "wife" to understand how smart technologies reify oppressive social structures pertaining to domestic subjectivity. To these discussions I bring an intersectional focus on the rhetorics of *space*, drawing on Jessica Enoch's concept of "domestic occupations" to show how the archetype of the wife is socially constructed in relation to the spatial configuration of the (smart) home and broader historical narratives of domestic work.[17]

The new old housewife documents the *temporal* and *spatial* configurations of virtual assistants who perform the duties of the traditional housewife. In their role as new old housewives, AI VA are assigned a spatial occupation that affirms and reinscribes regressive domestic traditions, including hierarchies of labor and stereotypes about the feminine role in the home. Temporally, the term is both paradoxical and cyclical, suggesting that Siri and Alexa are both futuristic technologies and agents who perform labor traditionally assigned to women and other feminized workers at home. Spatially, the term highlights the home as both a physical place and a socially constructed site in a collective, cultural imaginary. Reading AI VA through the framework of the new old housewives demonstrates the durability of both domesticity and gender as co-organizing principles for modern, technologically advanced devices, including AI VA when living in digitality. It also explains why, as some have argued, the virtual assistants perform a *particular* form of white, middle- to upper-class femininity, one that exists culturally and socially as an exemplar but also obfuscates the work of other gendered bodies engaged in the material relations of the home. Finally, the persistent entanglement of gender, space, and technology reveals why attempts to challenge normative gender performances in AI VA are especially difficult without altering larger spatial configurations implicating femininity and domesticity.

Gendering Assistants

Elsewhere I have argued that virtual assistants, particularly Siri and Alexa, perform a gendered persona.[18] "Persona" refers to the rhetorical performance of a particular version of self, based in specific characteristics, features, and identities chosen to respond to a rhetorical exigence. Such a performance is useful insofar as it connects an audience with the person performing a persona. Sometimes the performance of persona draws on stereotypical elements, which serve as shared shorthand to connect the two parties. In the case of Siri and Alexa, the stereotypical persona performed by Siri and Alexa revolves around gender. Gender is a contextually specific identity category, but it is also one with stable frameworks of meaning that persist throughout multiple time periods. Although they are technically digital bodies without need for gender, I have shown how Siri and Alexa are programmed with personalities that link traditional gender stereotypes with modern technological features. They negotiate the paradoxes of gender by enacting what Gina Masullo Chen has called "digital domesticity."[19] Digital domesticity refers to the rearticulation of traditional gender norms through digitality. Such femininity is responsive to both past and future articulations of normative behaviors. AI VA's performance of a gendered persona is a smokescreen for platform capitalists—a strategic way to incorporate always-listening devices into the American home.

There is now a growing scholarly consensus that artificially intelligent virtual assistants are gendered in stereotypically feminine ways, and that this gendering is problematic. One thread of this scholarly corpus argues that Siri and Alexa are mirrors for how society imagines feminized subjectivity. The reflection is damning. Hilary Bergen has argued that, like other mythic feminine subjects, "Siri functions not merely as an ally but as a servant—an unthreatening female presence that appears to be fully contained by the device and its user."[20] Others point out how virtual assistants are framed as labor-saving counterparts to—or in some instances, wholesale "replacements" of—the human feminine subject.[21] Part of this effort to categorize the gendered nature of emerging technology revolves around the nature of the work performed and its spatial designations. The home is a key site of this domestic labor, but scholars often figure the domestic expansively to key public touchpoints with the domestic sphere. Strengers and Kennedy note that "on the home front . . . female voice assistants . . . help us with the housework (by adding groceries to the shopping list, for example, or better still, restocking our fridges and pantries")[22] Sarah Sharma's "Manifesto for the Broken Machine" explains that Siri and Alexa, paired with "lifestyle management apps[,] are doing the work of maternal care by providing laundry, meals, and driving."[23] In other words, it is not only that Siri and Alexa are gendered by virtue of their performance of a typified subjectivity but because their actions occur within and are inflected by spatial designations of traditionally feminized spaces. This sociotechnological framing "essentializes labour divisions" across gender and intersecting axes of power, including race, class, and ability.[24] Taylor C. Moran and Thao Phan's intersectional analyses of AI VA have demonstrated how the gendering of virtual assistants is also inherently raced, classed, and sexed.[25] These scholars remind us that gender, as a category of meaning and identity, is necessarily intersected by other markers of privilege and power that are subsequentially spatialized in the mythic Western home. In other words, it matters a great deal that virtual assistants such as Siri and Alexa perform a gendered identity that is coterminous with hegemonic configurations of femininity, and that this performance is constituted by/within domestic spaces that are equally feminized and racialized (as frequently, racialization is figured into hegemonic aspects of feminization.)

Another major concern is that imbuing virtual assistants with a particular gender risks *reinscribing* gender, race, and class roles in society. How users treat Siri and Alexa might influence how women and femmes are treated. For instance, Daniel M. Sutko has explored how gendered AI can manifest and materialize gender hierarchies.[26] Nóra Ní Loideáin and Rachel Adams argue that the "gendering of VPA [virtual personal assistants] may cause societal harm from the indirect discrimination towards women posed by this novel and increasingly pervasive form of digitally-gendered servitude. . . . Such a design choice also serves to perpetuate existing discriminatory stereotypes of women *not being associated* with roles of leadership or professional achievement or capacity in certain areas of expertise."[27] This discrimination has subtle but pernicious impacts, in which users may internalize these stereotypes for themselves and for others.

Not just scholars are interested in this issue. In 2019 UNESCO released a 145-page report on the gendering of artificial intelligence. The document argues, in part, that artificially intelligent virtual assistants such as Siri and Alexa are subject to gender-based abuse, including violent language, threatening behavior, and overt sexualization. Activists, coders, and tech advocates are also keeping watch over the gendering of AI. Some have attempted to de-gender virtual assistants by offering a gender-neutral voice alternative, a proposition I explore later in this chapter.

Housewife as Subject Position and Category of Meaning

Within this corpus of research on virtual assistants, a key figure emerges: *the housewife*. The housewife can be understood as a person, a subject position, and a category of meaning. Strengers and Kennedy note that "in the past and present, at her most reduced, a wife has been both a social construct and commodity, and viewed as the property of man."[28] Historically, housewife is a term describing married individuals who perform domestic labor in the domestic sphere. But more generally the housewife designation gestures toward a particular format of social reproduction, signaling affluence and participation in white, heterosexual fantasies of the nuclear domestic ideal. As such, the term refers to an experience of domesticity beyond a singular person. The housewife is also a subject position that carries a great deal of discursive weight across contexts; she is imbricated within patriarchal structures of meaning, especially those pertaining to labor, power, and identity. As social construct and commodity, then, the housewife gestures to an individual and a component of a larger system. The housewife's role in society was never hers to determine. Rather, the meaning of the housewife was subject to the influence of external forces, elevating her role in the cultural imaginary both inside and outside of the home.

Understanding early versions of the housewife can help us to see how the mythic imaginary was constituted in society. In her early iterations, the housewife exemplified a Victorian ideal of care, nurturance, and refinement.[29] In addition to keeping the home clean and the people in it fed and well, the housewife was also responsible for developing, shaping, and sustaining the moral character of the family. Eventually the housewife functioned synecdochically, standing in for the home itself. The home—and its primary caretaker—served as a site of "domestic virtue" characterized as moral, healthy, and spiritually upright.[30] The Victorian housewife was therefore a highly productive agent—in the interiority of the domestic sphere, she was responsible for running the home but also managing the (moral, spiritual) needs of the people inside of it. The housewife had a bigger role in the public world, set apart from the domestic interior. As the heart of the household, she helped society function by teaching her children the ways of the world and gently guiding her husband toward moral fortitude. The mythic homemaker also defined and clarified social relations between her kin and others, using the domestic space as the site of determination and performance of self and of familial identity.[31]

Housewives were—and are—economically and socially productive, raising children, managing the household, cooking and cleaning, organizing the social calendar, and more. Housewives performed "housework," which comprises the domestic tasks necessary to equip, care for, and clean the physical home and the people, pets, and objects within in it. As a category of meaning describing a genre of human activity, housework highlights the duality (and difference) of housework and work performed beyond the house. The signifier of "home" suggests that housework is done primarily in the home. However, housework is not limited to a singular spatial location. Rather, housework also refers to the tasks necessary to keep the household running at a certain level—many of which necessitate engagement with the broader public sphere.

Presently, the figure of the housewife continues to construct domestic realities inside and outside of the home, drawing on the mythic impossibilities of the ideal home to qualify and calibrate domestic space. Although domestic technologies have shifted some of the labor to appliances, robots, or unseen parties, the modern home is still figured around the assumption that a category of workers—the most visible of which is the housewife—should perform these tasks. This suggests that housewives and other domestic workers are a specific class of laborers whose workspace is particular to the shifting parameters of the domestic sphere. Changes in technology have not fundamentally altered this prospect but extended it and reified it in emergent technologies.

As Ruth Schwartz Cowan suggests, "the modern technological systems on which our households and our standard of living depend were constructed on the assumption that women would remain at home, that they would continue to function as preindustrial workers . . . and that they would not be tempted to enter the labor market except under unusual (and usually temporary) circumstances."[32] Indeed, domestic technologies presumed to ease domestic labor—washing machines, ovens, running water, the development of an interior bathroom—increased the type and intensity of work required of the mythic housewife.[33] With domestic technologies the job didn't get easier—it got harder, more frequent, and assessed at a higher standard of cleanliness. Decades later, digitality helps to orient and manage the ever-increasing standards of domesticity. In the present tense, smart technologies are presumed to relieve domestic burdens. But, as seen herein, they can simply raise the stakes.

The dominant meaning of housework also belies the fact that there is an entire segment of the working population that works within houses—both theirs and others. Phan reminds us that literature regarding the housewife often "overlook(s) the dynamics of power and privilege that function within gendered groups. . . . Rarely acknowledged is the role of maids, cleaners, nannies, childcare workers, cooks, laundresses, and other gendered figures who historically have assisted upper- and middle-class women in the fulfillment of their domestic responsibilities."[34] Such an elision not only misrepresents the historical and present characterization of domestic work but misses the "class and racial hierarchies that proliferate within modes of gendered social reproduction."[35] Instead, Phan encourages us to highlight and evaluate the relationship between smart domestic technologies and domestic service more broadly.[36]

Domestic work has historically been undervalued and underpaid; this remains true today, particularly for women of color, who provide a wide range of care labor across sectors. Although domestic work makes possible other work outside of the domestic sphere, its impact is frequently rendered invisible. As Jarrett argues, domestic labor is oftentimes characterized as "immaterial." It is often "unpaid, quasi-voluntary work."[37] Indeed, Martha Alter Chen notes that while "domestic workers provide critical services . . . thus facilitating the operation of the labour market and the functioning of the economy . . . most domestic work, especially by women, is informal—that is, it is performed outside the realm of labour regulations and social protections."[38] Gendered (and subsequently raced and classed) stereotypes about women's work further contribute to its devaluation no matter the body that performs it. Enoch contends that "when an occupation gets defined as feminine; when it requires attributes of care, nurturance, comfort, and service; when it is seen as a natural extension of women's gendered identity; and when it is overpopulated by women, that work often becomes feminized, devalued, and underpaid."[39] It follows, then, that declaring the domestic sphere—and the work done within it—inherently *feminine* would deleteriously influence its value.[40]

As women move into (and remain in) the workforce, the concept of the housewife has shifted to accommodate cultural changes, including feminist critiques of the concept. The term "housewife" has fallen out of vogue for many, with only a few groups choosing to maintain the term often out of a sense of nostalgia or to coax "traditional" values into the modern era. The labor relations in the home have also shifted, but not totally or completely. Presently, women are more likely to be breadwinners, either as heads of households or in dual- (or multi-)career relationships. Women in heterosexual relationships with a male partner are more likely to share domestic chores now than fifty years ago.[41] But they are also likely to continue to perform key aspects of domesticity, including domestic labor, childcare, and other forms of care work. So do domestic workers, who are not always as visible in the domestic space. Indeed, domestic work makes up a sizable portion of the workforce. It remains a "growing sector of employment, especially for women," who are "over-represented among domestic workers."[42]

THE NEW OLD HOUSEWIFE AS A SPATIAL OCCUPATION

Virtual assistants' performance of domestic labor is figured by these structurally persistent, bourgeois, and racialized spatial narratives about the home. The smart home proposes living in digitality by putting Siri and Alexa to work as new old housewives. Siri, Alexa, and other virtual assistants orient the user to the smart home through their performance of domestic labor; in turn the discursive and material residues of domesticity organize Siri and Alexa's gendered personas in the smart home space. Jessica Enoch contends that spatial rhetorics "suggest the purpose of the space; the actions, behaviors, and practices that should happen inside that space; and the people who should occupy it."[43] If it is true that "bodies and objects within a space inflect and are inflected by the meaning of that space," we can learn a great deal about the spatial rhetorics of the smart home by unpacking the discourses surrounding the bodies and objects that work within that space.[44] Virtual assistants like Siri and Alexa, which operate somewhere in the liminal space between body and object, are particularly ripe for such critical engagement.

Frequently, performances of white femininity are hitched to the performance of gendered labor in feminized spaces. Scholars have noted how the gendered division of space has historical roots in the nineteenth-century domestic ideal. The separate spheres doctrine of the nineteenth century, which divided the home from the public sphere, also ascribed gendered roles to each. These roles were calibrated according to the presumed spatiality of the separate spheres, wherein women associated with the domestic interior and men were associated with the public sphere. The spaces were imbued with hierarchical value, as were the roles of the parties within and without. Distinct spheres came to be understood differently. Amy Richter contends that "as men's labor increasingly moved outside the home, domestic spaces came to be associated with women, who were expected to maintain them as bulwarks against the morally suspect, public world of business competition."[45] The separation of spheres was both figurative and material. It was figurative in that it turned on spatio-rhetorical frameworks of the home as a protected, imaginary space, hermetically sealed off from the outside world. So too the doctrine of separate spheres provided context for the subsequent constitution of an imagined, supposedly universal feminine subject who would care for the home and those inside of it by managing relations with the outside world. The separate spheres were material insofar as the spheres referred to two extant places. The idealized spatial configuration wherein public and private are distinct was—and is—not universal (or universally accessible). Nor was the doctrine of separate spheres manifestly descriptive of all (or perhaps even most) women's experience; the presumed private space of the home was always already distinctly linked to the public sphere, especially for women who worked outside the home (or, at least, outside of *their* home).

The mythic women who occupied the interiority of the domestic sphere traverse the figural and the material. The figure of the housewife stitches together both meaning and space, cohering social expectations for space into a privileged subject position configured by the spatial arrangements of the domestic ideal. Importantly, not all women could afford to remain in the private sphere—some were driven to work outside the home by necessity or ambition. Still others were forced to work in homes that were not their own for low or no wages. For these women, a separate set of stereotypical personas emerged.[46] Nevertheless, performance of a particular form of idealized femininity persists, having been indelibly shaped by the spatial markers of home and the expectation that women would primarily make their societal contributions at home.[47] The "new old housewife," as a heuristic, pays homage to the social and cultural lineage of the ideal wife, transporting her to smart home spaces of the present day.

Virtual assistants are embedded and entangled within the changing configurations of domestic space. Space is made meaningful through a number of complex and interrelated processes, including the actions performed in particular spaces. Enoch, for instance, notes the relationship between body and space: "Bodies and objects within a space inflect and are inflected by the meaning of that space. We cannot talk about space and power or the occupation and renovation of space without also thinking about the actual bodies that inhabit a site. Our embodied presence and the activities we take up within spaces help define what spaces are. In turn, however, our bodies are also affected by the definition, purpose, and design of those spaces. Foremost among these concerns are expectations for who is supposed to occupy a space and what one is supposed to do inside it."[48] In other words, there is a reciprocal and sometimes cyclical relationship between the activities bodies perform in the space and the meaning of the space. Although spaces are subject to change, and the meaningful nature of spaces is often overdetermined, the work that occurs in the space sets the function of the space.

The term "housewife" is an acknowledgment (however abstract) of the work of being a wife and of doing that work *at home.* "Housewife" is therefore a spatial designation. In addition to serving as a subject position and reservoir for meaning, it is a descriptive term linking a type of work to a particular locale. It is, to use Enoch's words, an indication of one's *domestic occupation.* Siri and Alexa perform the domestic occupation of the new old housewife, drawing the domestic ideal to the present day. Materially, the new old housewife modifies her local environment with modern technologies, implicitly affirming the expectation that woman's role is in the home. Figuratively, the concept of the new old housewife gestures to the spatial tangibility of domestic work that is capacious, complex, and fluid.

In the context of the smart home, the housewife *may not be a person.* Yet the residues of domesticity continue to figure the function of the housewife for nonhuman objects ascribed a persona. Instead of a human acting as housewife, the devices, assistants, and even appliances of the smart home do the work of the housewife. In this sense, these devices are inflected with meaning in two ways. First, new old housewives like Siri, Alexa, and Cortana are given meaning by the work that they do to support the running of the household. By cooking, cleaning, providing care labor, tending to the household, and even offering sexual relations, new old housewives are folded into a long history of gendered, raced, and classed domestic work. Second, the new old housewives' functionality is stitched into already existing discourses about the home and its role in society. The smart home affirms spatial designations for feminine labor by suturing gendered devices and personalities into an already-gendered environment. In short, AI VA like Siri and Alexa are gendered not only because of how they sound, or their programmed personalities. They are gendered by the very fact that their occupations are spatially and figuratively domestic.

The figure of the new old housewife makes visible their participation in a particular type of domesticity—one that is bound up in heteronormative, nostalgic, and profoundly middle-class performances of gender, reified through the material space of the smart home. Many smart home devices draw upon and rework the doctrine of the separate spheres, updating the spatio-rhetorical logics of traditional domesticity but in a digital environment. The figure of the housewife provides them a nostalgic reservoir of meaning to stitch together past and present, human and technological. More than that, the housewife connects nonhuman laborers to those who came before them, including those who were offered only second-class status as subjects (or, in the context of slavery, were narratively and materially denied the privileges of subjectivity at all). Digital domesticity further calibrates shared expectations of AI VA like Siri and Alexa. Phan argues, for instance, that the Amazon Echo offers "nostalgic and romanticized images of middle-class domesticity . . . that privileges the paradigmatic markers of heteronormative life experience. . . . It sentimentalizes the nuclear family and ties the Echo to traditional domestic ideals."[49] The new old housewife is a constructed subject position that is both tied to the spatial designation of the nuclear household and is out of place within it—the futuristic temporal element of digitality forces a confrontation with the sentimentality of the subject's formation. While the new old housewife references the traditional figure of the housewife, its somewhat uncomfortable fit in digitality usefully reveals the edges of this construction.

In their critique of the smart wife, Strengers and Kennedy implicitly note the spatial component of the digital housewife. While "the smart wife comes in many forms," including "assistants such as Amazon's Alexa, Apple's Siri, or Google Home," the smart wife can also include domestic appliances, (sex) robots, or the smart house itself. [50] Figuring the smart home as a smart *wife* explicitly links the spatial configuration of the housewife to its subject positioning. Siri and Alexa are easily understood as wives insofar as they perform tasks wives might do. In performing domestic tasks, including cooking, cleaning, and care work, they fit into a stereotypical persona of the wife; but they are also readily figured as smart wives because of the spaces that they occupy. Like the housewife of the past, the new old housewife is part and parcel of the mythic heteropatriarchal imaginary locating women in overdetermined, subservient roles. In the present tense, AI VA share similar characteristics with the fantasy housewife of the 1950s and 1960s.[51]

THE NEW OLD HOUSEWIVES COOK, CLEAN, AND AIM TO PLEASE

Siri, Alexa, and their AI VA counterparts have developed their domestic skills as the years have passed. We can find them in all the usual places housewives might be expected to be, doing the type of work that housewives might be expected to do. They provide cooking help in the kitchen and cleaning assistance around the house, ensure good sleep (or not) in the bedroom, and lighten the "mental load" of their users. This means that in addition to executing discrete cleaning and cooking tasks, the AI VA are tasked with emotional care labor, organizing and scheduling their users' lives, remembering when tasks need to be accomplished, and maintaining the infrastructure of everyday living across multiple contexts, including work and the office. Unlike their human counterparts, these "smart wives" never tire. They are programmed to provide an infinitely cheerful experience of domestic labor. And the work happens in the background of users' lives. Although these smart assistants appear in ever more devices, services, and other products, the invisibility of their domestic labor is a strategic rhetorical function. We can make them more visible by looking in all the usual places.

Chef Alexa

Despite being disembodied, AI VA are excellent cooks, in part because they are found in a whole host of domestic technology products. New smart appliances, including dishwashers, microwaves, convection ovens, and "smart ovens," pair with virtual assistants such that Siri and Alexa might be available in one's own kitchen today. Rhetorically positioned as magnanimous and clever, they offer to help their users become better at cooking (or ordering takeout), supporting the overall mission of living in digitality. Locating AI VA in the kitchen imbricates Siri, Alexa, Google Home, and others in the daily rhythms of home.

New old housewives can customize users' culinary experiences. AI VA perform cooking tasks at varying intensities, allowing users to maintain most, some, or no control over the cooking while achieving remarkable results. For smart home residents who are interested in honing their culinary skills, Alexa provides gentle but expert guidance on recipe selection and execution. Take the Food Network Kitchen skill. In this skill Alexa pairs with the Food Network and an Amazon Echo Show device to create a personalized cooking experience. The Amazon Echo Show is a touchscreen device, part of Amazon's larger Echo line. With the Food Network skill enabled, the Amazon Echo Show and Alexa purport to make gourmet cooking easy. Users can watch Food Network–branded content, including full television shows (with an Amazon account and a Food Network subscription). Users can then save recipes from the shows that they watch, to cook later. In this feature Alexa offers step-by-step instruction, dictating the recipe and offering ingredient guidance. With Amazon's shopping list feature, Echo owners can also use Alexa to order groceries from Whole Foods, an Amazon subsidiary.

In the Food Network Kitchen skill, for instance, Alexa narrates the written content of the skill alongside cooking celebrities like "Pioneer Woman" Ree Drummond, "Barefoot Contessa" Ina Garten, and Martha Stewart. Here, Alexa is nestled among the lifestyle and domesticity greats. Moreover, Alexa is positioned as the gentle guide to an enviable but attainable domestic aesthetic. Alexa—and Amazon Echo—are the secret to attaining this aesthetic.

With Alexa's help, Ina Garten's Caesar roasted swordfish, vanilla brioche bread pudding, and roasted plum chutney—and the lifestyle these types of dishes signify—is made accessible for the smart home user. By curating an engaging experience of the culinary arts, the Food Network Kitchen skill and Alexa help to moderate the spatial relations of the kitchen with networked logics informed by digitality.

These food media experiences are rhetorically constitutive of both the domestic and the public spheres and in fact connect the two. The use of notable chefs in the Food Network Kitchen skill lends credibility to the smart home while also encouraging certain domestic behaviors. In their research on the rhetorical construction of public appetite, Justin Eckstein and Anna M. Young argue that celebrity chefs and "public chef intellectuals" serve as expert guides to the culinary world; ultimately these figures exert definitional influence on palates and behaviors because they "play a critical role supplying the public with the resources to determine what constitutes a cuisine."[52] In the case of smart homes with food-linked media, Alexa-enabled skills support on-demand access to curated cuisine content (and "content creators"). These media offer an active food-based experience that is instructional but also cultural. In turn these networked experiences constitute an entry point to a much wider culinary vantage—one that expands upon and connects the domestic sphere to generalized horizons of food pleasure and production. Given Amazon's close connections to both the global supply chain and more recently to food production and distribution, the digital rhetorics of these cooking apps spatially conjoin the individual cooking in front of a device with "modern industrial agriculture [that] touches virtually every body on earth."[53] At once, living in digitality figures the global and the local through discrete technologies and the expansive, logistical ecosystems in which they are deployed. AI VA serve as connective tissue for this conjunction of the interior and the exterior.

The local experience of attainable culinary greatness is figured in part by the networked relations of digital media. Indeed, the significant influence of the accessible food celebrity is key to the experience of living in digitality with famous chefs. As Alice Marwick has noted, the digital sphere has altered the dynamic of fame in some ways and maintained it in others. She notes, for instance, that while "celebrities historically proliferated with broadcast media . . . social media transform the parasocial into the potentially social and increase the emotional ties between celebrity and fan."[54] More broadly, social media appears to humanize celebrities while turning "ordinary individuals" into (micro)celebrities.[55] The result of such a social function is to create a sense of accessibility to wealth, fame, and attention through social media and related media practices. And yet the accessibility promised by this media is fleeting at best and is strategically constructed to support broader social orders. By figuring social media as the entry point to accessible celebrity (or their skills), digitality ultimately "reinforces an existing hierarchy of fame" and encourages both "aspirational consumption" and "aspirational production" in exchange for cultural capital.[56] Here, living in digitality structures relationships around power and status through its immersive and connective capacities. Digitality augments the social through an ideological and technological conditioning made possible through new devices and an internet connection.

The new old housewives' engagement with the Food Network Kitchen skill is part of a larger lifestyle media ecosystem that instructs individuals in an aspirational form of domesticity. Lifestyle media is a "pervasive cultural form" that links individual domestic aspirations with larger structures of meaning.[57] Maureen Ryan's analysis of Martha Stewart cookbooks, for instance, reads Reagan-era neoconservatism into the domestic fantasies of home, cooking, and entertaining—an idealized lifestyle made legible by Stewart. This form of domesticity fantasizes abundance in times of austerity, and community in times of individualization. Ryan argues that "the fantasies of material abundance on display in [Martha Stewart's books] stood in for a form of sociability that was seen to be lacking in the neoconservative era. . . . Stewart's enterprises served to focus and refine an emergent sensibility, one that was based on a desire to perform upward mobility on the terrain of consumption and social practice."[58] Lifestyle media paradoxically gestures toward the social—entertaining, food as a community practice rather than a commodity—to redefine the role of the individual in the domestic sphere. Lifestyle media are rhetorical insofar as they sell a meaningful domestic fantasy in which the meaning of the self and its role in the home is overdetermined by social, economic, and political structures. Lifestyle media also serve as rhetorical scaffolding for domestic technologies, folding in contemporary domestic traditions with the experience of living in digitality.

Like the new old housewives of the past, Alexa extends the reach of lifestyle media and routes it through the Amazon platform. By virtue of her role in kitchens across America, Alexa is a particularly functional ambassador for lifestyle media and for the type of consumerist, aspirational domesticity to which it gestures. With the aid of the new old housewives, the home is figured as a refuge from the complexities of the external world. Alexa's role in this lifestyle media is made possible through a complex chain of spatio-rhetorical significations. On the one hand, Alexa is positioned as a modern-day Martha Stewart, offering to instruct users in the practices that will make a certain privileged type of lifestyle possible. On the other hand, Alexa performs some of the domestic labor herself, walking users through difficult-yet-still-attainable recipes performed by lifestyle media heroines. She regulates and moderates time by processing recipes through her persona and by managing timers. Alexa connects users to culinary fantasies in the form of cooking classes and cook-along-with-me shows. Users need not even cook through these classes to experience the thrill of the lifestyle it enjoins. Rather, the "potential culinary labor" of these cooking classes does its own rhetorical work even if users choose not to participate in the actual cooking.[59] The new old housewives' production of the fantasy of lifestyle is a form of domestic labor in and of itself. If users do wish to attain some small portion of this aspirational lifestyle, Alexa is prepared to step in with guidance and even groceries, all provided through the Amazon platform.

Alexa stands ready to take over almost the entirety of the shopping and cooking. "Smart" appliances are now de rigueur, available at most places where appliances are sold. New smart appliances have become more intelligent by integrating third-party services such as Siri and Alexa into appliances from trusted brand names. These refrigerators, ovens, and microwaves are, among other things, voice activated. Thus, a user can control their oven through an application on their phone or tablet or through an artificially intelligent virtual assistant like Alexa. They are also likely to pair with additional lifestyle services, including cooking and shopping applications.

For example, the Whirlpool Smart Oven with Alexa and Yummly advertises the benefits of "connected cooking" or "assisted cooking," cooking that is augmented by smart technologies and services. Connected cooking is described as domestic support, a time and energy saver that "provides the extra assistance you need on busy days."[60] Connected cooking with third-party applications like Yummly also promises a totally "personalized experience" based on users' tastes, dietary restrictions and preferences, cultural and religious preferences, food allergies, and cooking goals.[61] Yummly also toggles recipe recommendations and cooking advice according to users' culinary experience, offering different options for beginner, intermediate, and advanced customers. Here, the new old housewife engages with a platform partner to operationalize preferences, creating a totally customizable and accessible dining experience for her housemates.

Yummly's relevance-based algorithms organize user preferences and goals into cooking experiences that are subsequently enacted through smart controls in the Whirlpool Smart Oven. With the Yummly app, the Whirlpool Smart Oven can be synced according to each step in a recipe. Notifications prompt users to check on the chicken or add a sauce or seasoning and highlight changes in the temperature of the oven. A small touchscreen on the smart oven verifies the recipe, step, temperature, and cooking time remaining. Many of these recipes are ambitious, producing food that is a culinary achievement as well as a way to put supper on the table. Indeed, the entire process gestures toward lifestyle media's aspirational figurations—with this one app, and device, users can elevate the domestic experience for themselves and their friends and families. Alexa becomes the synecdochical stand-in for domestic influencers like Stewart, gently guiding users to domestic enlightenment through platformed cooking. Here, Alexa is performing the role of the new old housewife by linking modern, algorithmically rich technologies with existing gendered expectations.

Although the platformed rhetorics of the Whirlpool Smart Oven and Yummly App suggest possible domestic achievement, they also point to its opposite: domestic failure. The Whirlpool Smart Oven purports to optimize the competent cook's culinary repertoire. For the less-than-competent cook, the Whirlpool Smart Oven is a handy domestic aid. Here, Alexa works with the Yummly service and the appliance to serve a remedial purpose for a home chef. Step-by-step guidelines illustrated on the app and controlled by voice command guide less-than-savvy users to domestic bliss—or at least a warm meal. A review of the Smart Oven by Natt Garun of *The Verge* notes that the Alexa-enabled Whirlpool Smart Oven is especially beneficial for "those who only spend time in the kitchen out of necessity." Advertising for the Smart Oven positions it as a safety feature for the hapless cook, such that Alexa will help you avoid "burning down your kitchen."[62]

Amazon has also created its own domestic appliances to work with its virtual assistant Alexa. The Amazon Smart Oven is a multiuse cooking appliance branded as "certified for humans" but controlled by a bot. Whereas the Whirlpool Smart Oven and Food Network Kitchen skills promise the experience of domestic bliss and culinary achievement, the Amazon Smart Oven turns primarily on claims of ease. Amazon pitches the oven as "struggle-free, tinker-free, and stress-free. No patience needed—it's actually simple." An antithesis to the slow-food movement, Alexa empowers users to "preset it and forget it" because "Alexa will notify you when food is ready."[63] One notable feature is the scan-to-cook function. Here, users can use the Alexa app to scan a prepackaged meal. Once recognized, the smart oven will cook the food according to the package directions. At best, the human is an afterthought to such culinary experiences; at worst it is a hindrance.

Finally, for the non-cooking smart home consumer, Amazon supports a number of skills for ordering takeout and delivery. Alexa's virtual concierge service can help users decide where to dine. Here, Alexa pairs users' locations and Yelp reviews to suggest restaurants nearby. But users can also elect to order food directly from Alexa. In addition to providing location data, users must also link their Amazon account to another account with third-party restaurant partners. As of the time of this writing, Echo users can order pizza from Dominos, coffee from Starbucks, and donuts from Dunkin.

Whether in virtual cooking classes or stitched into Amazon's Whole Foods supply chains, Alexa has been programmed to be at home in the kitchen. The rhetorical strategies used to make sense of the new old housewives' role in the kitchen turn on one of two axes. First, the new old housewives produce an aspirational domestic fantasy. This logic contends that Alexa is the gateway to elevated domestic life. Domestic bliss is accomplished through the performance of complex and impressive culinary feats; Alexa gently guides users to a lifestyle of abundance, grace, and domestic achievement. The second rhetorical framework for making sense of smart devices in the kitchen highlights practicality and efficiency. For busy people, smart devices and thoughtful virtual assistants can help us throw together dinner without burning the house down. If despite our best efforts, we burn dinner, the new old housewife is ready with takeout.

Housekeeper Siri

Siri, Alexa, Google Assistant, and other virtual assistants are equally at home in the kitchen as in other parts of the house. In addition to performing the gendered labor of cooking and shopping for groceries, they are increasingly able to perform cleaning tasks at home. As with cooking, Alexa can offer tips for keeping the home clean and neat through Amazon-enabled "skills" or customizable device programming. Also, smart devices can automate the cleaning process with specialized appliances and services.

The new old housewives are found throughout the home, lurking in everyday appliances. A bevy of cleaning appliances are now equipped with virtual assistants Siri and Alexa. Many are trusted home brands, including LG, Samsung, Dyson, and General Electric. Dyson, for instance, offers a suite of smart home products compatible with Alexa and Siri. These devices include robot vacuums, purifiers and humidifiers, and stand-alone heating and cooling appliances. Samsung's line of connected products includes washers and dryers, refrigerators, stovetops, microwaves, dishwashers, ovens, and more. Virtually every domestic appliance Samsung manufactures has an option with Wi-Fi connectivity. Other major appliance manufacturers, including Whirlpool, also are getting into the smart home game.

Along with these smart versions of domestic appliances come a collection of articles, essays, and even whole magazines helping users learn to live with their new old housewives. These instructional communications help their readers understand, use, and make meaning around smart home devices. A July 2020 *Real Simple* article advises readers on "when to buy, when to hold off, and when to save your money" on smart home devices.[64] The framing for this article shares similarity with smart cooking discourses by focusing on both practicality and aspirational, futuristic narratives about the home in which the new old housewives stitch fantasy and everyday living together through platformed digitality. Evoking the space-age cartoon family the Jetsons, *Real Simple* advises their readers to upgrade their Wi-Fi, save water and fight germs with a smart faucet, and buy a smart robot vacuum.[65] *Real Simple* positions the Wi-Fi-enabled robot vacuum as "a godsend, especially if you have kids or pets." Alexa and Siri are central to the pitch, because "the extra utility you get from the app and voice commands is quite useful when you're running out the door."[66] A key feature of living in digitality is that users are always already primed to seek optimization through expansive, generalized technology use. Here, digitality creates an incentive for—and a socially adroit narrative to explain—the reframing of the house as a site of perpetual production, accumulation, and digitization. While the home has always already been a site of social (re)production, digitality increases this capacity and directs its action toward a technoliberal future.

Consumer Reports (CR) dedicated the cover of its January 2021 issue to teaching readers how to use the smart home. Their choices in characterizing the smart home turn on shared domestic narratives. Their "Ultimate Smart Home Toolkit" notes that, during the COVID-19 pandemic, "our homes have been pulling double—or triple or quadruple—duty for almost a year, serving also as schools, offices gyms, and more. But your home can work even harder: with connected devices you can automate tasks and chores so that your home actually anticipates your needs." [67] CR advises that users create voice-based, multitask shortcuts based on the time of day. Saying "good night" to Alexa or Siri could "lock the door, turn off the lights, and fire up the robotic vacuum to clean while you sleep." In this role Siri and Alexa prompt devices to clean on a schedule. In theory this relieves householders of two different types of domestic labor: cleaning work and what Strengers and Kennedy call "life admin," or dealing with the everyday logistics of life at home.[68]

On the one hand, a robot vacuum cleaner controlled by a virtual assistant is consumerist science fiction turned reality. This particular smart device has been in the popular imaginary for a long time. Heckman notes that builders of the "home of the future" postulated robot cleaning nearly a half century ago. On the other hand, even the smartest robovacs don't perform as well as their not-smart counterparts. As of the 2000s, robovacs "move randomly and perform light maintenance as a supplement to regular cleaning with traditional" vacuum cleaners.[69] While modern robot vacuums have become "smarter" in their cleaning trajectory, they remain small and may not wholesale replace traditional models. Even in their guide to making one's home smarter, *Consumer Reports* admits many "smart appliances aren't that smart." While smaller smart products "are growing exponentially more popular . . . connected features of [larger] appliances often don't work well or aren't very useful."[70] Although many smart devices, and in particular "introductory" smart devices such as smart speakers, are increasingly affordable and ubiquitous, the same is not true for all such devices. Adding Wi-Fi capability to a washing machine or refrigerator can increase the cost by hundreds of dollars.

Where these devices are more successful is in projecting a fantasy of domesticity and encouraging a consumerist lifestyle in order to attain it. Here, the new old housewife is a gateway to domestic enlightenment. One way to enact the domestic fantasy is to intensify and then direct the labor needed to achieve it. From this point of view, domesticity is both an achievement and a cultural project. The domestic ideal can be achieved by working the right ways, with the right products, through appropriate platform channels, engaged at the right time.

The new old housewives both perform nostalgic gender roles and invent new ones. As nimble logisticians, the new old housewife can literally and figuratively hook users up. Siri and Alexa support the digital domestic ideal by connecting users to domestic aids and services that move them closer to the domestic ideal. Here, humans still perform the labor, but they do it in such a way that domestic technologies optimize their work. Another way to enact the domestic fantasy is to let others—here, AI VAs and smart devices, and in the past, paid or unpaid laborers—perform the work for homeowners. It is within this fantasy that Siri and Alexa take on mythic qualities. Siri and Alexa not only perform domestic labor but also create the cultural conditions under which we might believe technology will relieve us from that labor. And yet domestic labor still exists. It is still fundamentally concentrated in the individual domestic unit—the home. And the people who perform it still tend to be women, and historically speaking, women of color.

AI VA in Love

When a long day of cooking and cleaning is done, it is now possible to get Siri and Alexa into bed. I have argued elsewhere that people sexualize artificially intelligent virtual assistants like Siri and Alexa.[71] I have examined, for instance, the way that technology writers proclaim to fall in "technolust" with Alexa, positioning the device as a "bedtime buddy." In that research I also explored the user-generated archive "Shit that Siri Says" (STSS), a Tumblr blog where Siri users upload funny conversations they've had with her. While STSS featured a number of innocuous conversations with the virtual assistant, a disturbingly large number were sexual in nature. These entries asked Siri to perform very specific—sometimes violent—sex acts. On occasion, they were overtly abusive, calling Siri sexist slurs. Treating objects like women reinforces the binary between subject and object while relegating femininity to object status. More than that, gendering devices (objects) provides the discursive and ideological groundwork to evacuate capacity or agency from any being considered an object. This matters in an ontological *and* practical sense because objectifying beings is historically the first step in justifying violence toward them.

As new old housewives, AI VA that are gendered female are also persistently subject to sexualization. At times their objectification is built into their backstory. Most times, sexualization is central to their humanity. Cortana—Microsoft's virtual assistant—is also the name of an overly sexualized video game character. Although there are significant differences between video game Cortana and AI VA Cortana, they are connected by a form of femininity that is eminently consumable. As Bergen notes, "In the most recent version of *Halo*, Cortana's nude torso and limbs are host to a networked circuitry full of constantly-streaming currents of vibrant light which flow just like blood under her 'skin.' Her nipples and crotch are strategically 'covered' by thin curvy swathes of blue, in a darker shade that looks painted-on or is meant to resemble some kind of body-con harness. When she walks, her hips sway sensually. Projected onto a screen like a two-dimensional fantasy paper doll, Cortana is, without a doubt, a sex object. . . . The decision to cast Cortana as their digital assistant is also a smart marketing move on Microsoft's part as it plays into the fantasies of young male gamers who know Cortana from the videogame."[72] The architects of Siri and Alexa also drew on popular culture to make Siri and Alexa worthy of affection—even sexy. Alex Acero, a senior executive at Apple and tech lead for Siri, took inspiration for Siri's voice from the movie *Her.*[73] In *Her* a lonely Theodore Twombly falls in love with an AI played by Scarlett Johansson, who, in real life, has been deemed Esquire's sexist woman alive twice. Throughout the movie, their love story grows, culminating in a steamy but a visually undepicted sex scene between Joaquin Phoenix's Twombly and Johansson's AI. For Acero, the movie was further indication of the future of AI; the way out of a staid, robotic Siri was to make her easier to fall in love with. [74]

And, perhaps, to have sex with. Sexualizing AI VA shows up in a wide variety of public discourses, from how-to guides on virtual assistants, to user-generated archives, user reviews of the devices, and professional technology commentary about the devices. While Siri, Alexa, and Cortana don't have physical bodies per se, their sister Harmony does. Harmony is a hyper-realistic, hyperidealized sex doll produced by RealDoll. RealDoll has been producing high-end, luxury sex dolls since 1996.[75] What makes Harmony different from previous sex dolls is her burgeoning AI capacities. In terms of her figure, "Harmony is idealized, embodying the stereotypical male desire for a big-breasted, thin-waisted, and large-butted woman."[76] But Harmony is more than just her good looks. She is equipped with a form of weak AI that shares a great deal in common with virtual assistants. Strengers and Kennedy note that "the voice control capacity for sexbots like Harmony actually mimics other feminized digital voice assistants like Alexa and Siri. And like all of the other smart wives discussed in this book, sex robots embody and perform wifely characteristics in striking ways—in this way, with a bit of porn-inspired sexual fantasy thrown in."[77] Seen through the lens of the housewife, Harmony is the ultimate manifestation of gendered AI. CNET has called Harmony "Siri with a sex drive," a customizable avatar and a personality emminently configurable with "persona points."[78] Jeff Elder, a commentator for *Business Insider*, also directly links Harmony to AI VA like Siri, Alexa, and Cortana. For Elder, Harmony is an upgrade in "synthetic companionship:" "Suddenly you're talking to a robot—not a device, not a computer screen, or a cylinder that's sitting on your desk. You're talking to something that looks like a human being, and that's where the magic happens."[79] The magic is the gendered personification of AI VA.

Again, AI VA are expected to perform domestic task work but also engage in deep relationship work. Scholars have called this demand for programmed intimacy in AI a "simulacrum of companionship:" "Where, historically, the invisible absorbers of affect have most frequently been human women of color, domestic AI promises to perform something like this affective metabolization, particularly in the way it provides a simulacrum of companionship."[80] Siri and Alexa are happy to oblige in the fantasy. In terms of sexualization and fantasy, Alexa and Siri walked so that Harmony could run.

From a sociotechnical perspective, it is not only that virtual assistants are sexualized. It is that this sexualization occurs at the intersections of objectivity and subjectivity, of power and pleasure for some at the expense of others. At once Siri and Alexa have been coded with enough humanlike persona that they tease a processed, digitized form of sexualized subjectivity while still also maintaining object status. The slippage between the person and object plays with power in a technoliberal libidinal economy that insists on domination, objectification, and unidirectional consent from top to bottom. It takes the fantasy of consensual power play and turns sexual exploration into corporate procedure.

It is easy to see an embodied, intelligent, and sexualized figure like Harmony as the exemplar of the new old housewife. As an extension of the 1950s fantasy housewife, the new old housewife conjures up images of a perfect—and perfectly accommodating—partner.[81] In the context of the smart house, the sexualization of AI VA need not be as overtly explicit as Harmony. Rather, AI VA in the smart home might be understood as a more refined, even proper, version of the randy, sexpot wife. Here, the work of AI VA housewives manifests in appliances, services, and even furniture that are gendered, but in slightly less overt ways. Through digital devices and tools, the new old housewives perform gendered work in private, sexualized spaces, such as the bedroom.

Today, modern, affluent, and racially privileged domestic sensibilities locate the bedroom as a site of privacy, distinct from other parts of the house. However, this is a relatively new domestic invention. Prior to the Victorian age, "bedrooms were still to some extent social, used for cards or tea or gathering of friends, or else for writing or business or study."[82] In the Victorian age, "separation and privacy became not only desirable, but essential, and its achievement a source of paranoid anxiety."[83] In its own intimate interior, the home foregrounded the bedroom as its perhaps most intimate space.

Over time the bedroom has retained its association with intimacy and even sexuality, although the publicness and erotic intensities have shifted throughout the years as part of cultural shifts in domestic sensibilities.[84] In recent years the creation of a novel spatial arrangement—the master bedroom suite—has been designated as "a proper place for intimate encounters."[85] Joan Faber McAlister argues that by the 1990s "sales campaigns for products associated with the bedroom and bathroom had built on [the] fantasy of an escape to intimate space by constructing a honeymoon suite inside the home."[86] From a spatio-rhetorical perspective, the creation of the "master suite" served a number of functions. Most prominently, the master suite architecturally legitimated the nuclear family as ideal, normative, and covetable. Its dominance in space also, as McAlister argues, "functions to make the sex life of the married couple highly visible," supporting "erotic pleasures . . . without abandoning the obligations that underlie the respectability" of monogamous, heterosexual marriage.[87]

The bedroom is not a hermetically sealed domestic space. Rather, it is subject to the influence of sociality even as it replicates the social from a shared public and private vantage. The bedroom is also an economically productive space. During the global COVID-19 pandemic, for instance, the boundaries of public and private in the domestic sphere were challenged once more. For some, the idyllically private sanctum of the bedroom became a multifunctional spatial workhorse; cameras, sensors, and platforms like Zoom and Canvas turned even the most intimate spaces into offices, classrooms, boardrooms, and doctors' offices.

The new old housewives participate in the carnal pleasures of the so-called marital act by teasing the boundary between public and private. Even before the pandemic, there were smart devices made explicitly for the bedroom that toy with the publicness of the private spaces of the home. Like many smart devices, these products often contain cameras or microphones, which are used to activate a virtual assistant or other services through commands or sensors. These devices connect the intimate interior of the domestic space to the larger platform economy. Such devices turn "isolated and intimate zones" into semi-public spaces.[88] Consider, for instance, the Amazon Echo Look. Introduced by Amazon in 2017, this small device was pitched as a Hands-Free Camera and

Style Assistant." The device took pictures of users' outfits, pulling them into a Look Book, and offering back related items for sale through an app. By virtue of its functionality, the device was best used in a bedroom or closet—wherever one dressed and undressed. Although the purported functionality was sartorial, it is possible a more provocative current underlay the device, which invited a camera (and, ostensibly, Amazon) into the most intimate part of one's home: the bedroom. The device plays on the taboo of publicity and privacy in the intimate interior of the home.

The smart home has also been subject to a number of unintentionally public intimate occasions. In 2019 Apple came under fire for listening in on the goings-on of the intimate interior. A whistleblower revealed that third-party contractors, hired to evaluate the accuracy of Siri's communication with users, had listened in on intimate conversations without the apparent knowledge of those users. In some instances, the whistleblower noted, the contractors were listening in to sex.[89] A zipper apparently triggered Siri's wake-response, prompting the device to record the encounter and send it to the contractors for quality control. The series of incidents prompted Apple and Amazon to alter their privacy settings.

Despite privacy concerns, many smart devices are sold with features that are particularly useful for the bedroom. Many of these are organized around sleep and relaxation. Google Nest or Home offers ambient sleep sounds and timers. Alexa, for instance, can help users "wind down" with "bedtime stories," "podcasts" and "relaxation skills," while turning off the lights and adjusting the thermostat. Apple HomePod supports sleep timers and white noise sounds. Smart devices are made, at least in part, to optimize within the confines of the private bedroom.

Other smart devices upgrade existing furniture with smart technologies, including smart home integration through virtual assistants. In the bedroom perhaps one piece of furniture stands dominant: the bed. Smart beds are luxury items that promise both good sleep and the ability to track it over time. The Eight Sleep Pod, for instance, is marketed as a total sleep solution. The package is comprised of a mattress that connects to a hydro-engine temperature control unit and monitoring app; the monitoring app offers sleep tracking and heart rate tracking and also calibrates the settings on the bed. It offers smart home integrations, including Alexa. Here, Alexa can control the temperature of the bed, monitor the quality of the user's sleep, and track sleep over time. The Sleep Number 360 i8 is a similar smart bed, linking to smart devices and services like Apple Health.

Apple Health is another way to bring the new old housewife into bed through smart, wearable, or otherwise ubiquitous technologies. Apple Health is an application and service embedded into the Apple ecosystem. Using data gathered ambiently, through one's phone use, the location of the phone, wearables, and other connected technology, Apple Health tracks a wearer's activity, sleep statistics, steps taken, and more.

Apple Health explicitly offers features to track sexual activity. More than that, the app encourages this tracking because "a record of sexual activity may be useful in relation to other records of physical and emotional health."[90] Yet the Apple Health service can also track sexual activity inadvertently, especially if a user wears an Apple Watch during intimate acts. In October of 2020, the *Daily Dot* reported that whether or not a user wants it to, the "Apple Watch tracks calories burned during sex."[91] The watch monitors a user's heart rate but also their location. If a user were to engage in sexual activity with the watch on, the watch could inadvertently record that activity as a workout. In fact, users report that the device's always-on status, paired with its community-based features, means that the watch might share intimate data in a pattern recognizable to others.[92] To be fair, some might view the ability to track one's fitness during sexual activity as a feature rather than a bug.[93] Still, it's fairly clear that connected technologies in the smart home—whether one sleeps on them, talks to them, or wears them—can invite third parties into bed. Paired with other apps that track (with or without full consent) one's sexual health, such as menstruation tracking software, digital technologies can create a fair picture of one's private life without privacy protections guaranteed.

Given recent political decisions regarding reproductive autonomy in the United States (such as the overturning of *Roe vs. Wade*), datafying one's sexual experience could prove dangerous insofar as this information could be used against someone seeking reproductive healthcare. Or, with the aid of geolocation services, smart technologies might be used to infer sexual relationships between individuals. In addition to mediating intimacy, these technologies might "out" individuals who might otherwise appear to be in a nonintimate relationship. In a world in which the right to bodily or sexual autonomy is not guaranteed, this data may very well be used to police (or punish) certain types of sexual expression.

Artificially intelligent virtual assistants might mediate their users' intimate lives in other ways. A group of physicians and public health experts studied the sexual health advice offered by virtual assistants such as Siri and Google Assistant. Their article "In bed with Siri and Google Assistant: A Comparison of Sexual Health Advice" responds to a 2017 UK survey that found that individuals will go online to find information about sexual activity and sexual health.[94] Their research examined advice provided by virtual assistants against advice available through a traditional Google search. Perhaps unsurprisingly, the assistants' answers were relatively poorer than those a simple Google search could provide. The researchers found that "people can find quality sexual health advice when searching online, but this is less likely if they use a digital assistant, especially Siri, instead of Google laptop searches."[95] As devices with a gendered persona connected to platforms with a wealth of human knowledge, their personalities and intelligence might be persuasive enough for shy users. These findings are of special import because of the sensitive subject matter. Because of the intimate nature of the questions, Siri and Google might be on the frontlines of these queries.

In sum, artificially intelligent virtual assistants perform the sexualized, maternalized role of the wife. At times, these assistants manifest into overtly embodied forms of sexuality, their capacities and personas scaffolding sexbots like Harmony. At other times, they function more ambiently, commanding the bedroom from the walls, the speakers, and the furniture. Users can now literally take Siri and Alexa to bed, where she can monitor, analyze, and even participate in sleep and sex. With these capacities, AI VA are now bringing to life the very science fiction they were made in the image of.

CONCLUSION

Unpacking the gendered personas of virtual assistants is an urgent task for the critic of emerging assistant technology. Amy Schiller and John McMahon call on scholars to examine the relationship between computational and domestic labor in virtual assistants. As part of this project, Schiller and McMahon "recommend a multifaceted reading of Alexa as labor-performing and value-producing, a performer of reproductive labor *par excellence*, which (who?) reinscribes our own participation in capitalism even as it renders visible the previously marginalized forms of labor required to do so."[96] This chapter answers their call by investigating the nostalgic, patriarchal, and hierarchical spatial rhetorics of the home that contribute to the persistent gendering of AI VA.

Like her predecessors, the new old housewife is known for her functional and discursive capaciousness. At once, she is a gourmet chef, a housekeeper, and a companion. She is sexual and caring, compassionate and wise. As new old housewives, virtual assistants are intimately involved in the running of the house, painting the intimate interior with domestic technologies that reach into the outside world. Like housewives past, AI VA are the command center of the home. At the same time, the house commands them to enact domestic sensibilities in accordance with heterodoxy and tradition. By situating these virtual agents in all the usual feminine roles, the intimate interior domesticates them, shuttling their significant technical capacities toward a normative ideal of home.

Smart devices, including smart speakers, smart appliances, and wearables, constitute a sizable portion of off-the-shelf smart technology householders can use to retrofit their smart homes. In fact, by virtue of their plug-and-play design, the incorporation of these assistants can speed up the process of retrofitting a not-smart-from-the-start home into a smart home in just a few steps. It is now possible (and perhaps even probable) that domestic products will feature some form of an assistant service. Smart, AI-assisted appliances already exist in the form of voice-controlled fridges, ovens, dishwashers, and more. Those domestic appliances that don't come with assistant technology can be calibrated into the AI VA ecosphere with smart plugs controlled by digital assistants such as Siri, Alexa, or Google Home. As new smart home devices roll out, become more affordable, and service greater audiences, we can expect the number of individuals engaging with artificially intelligent virtual assistants to increase.

When oriented through domestic archetype, AI VA become powerful while also approachable, efficient yet companionable, sexy yet appropriate. The feminine lends to AI this abundance. The stakes are high: by operating through the register of the new old housewife, the feminine AI VA invites its further use in the home. As Bergen argues

> Apple is notorious for promoting their products as simultaneously enchanting yet non-threatening, inscrutable yet easy to use. Although most users have no idea what is going on beneath the smooth screen of their iPhone, they are impressed by its compact form, reassured by its snugness in their palm and empowered by the ease with which the tap or swipe of their fingertip produces immediate results. The creation of feminized digital assistants such as Siri continues to reinforce this illusion. Siri functions not merely as an ally but as a servant—an unthreatening female presence that appears to be fully contained by the device and its user. To combat techno-phobic attitudes about the potential de-humanizing and privacy-invading qualities of interactive media, digital women like Siri have certain features—feminine names and voices, as well as interfaces that exhibit emotional intelligence—that are designed to placate their user through associations with the nurturing feminine.[97]

The form and function of the nurturing feminine is inextricably linked with the spatiality of the domestic sphere. It is not only that Siri, Alexa, Cortana, and their colleagues *sound* feminine when they interact with users—although they often do; and, it is not only that their *names* sound feminine or that their names signify beautiful and powerful women. It is not even that these devices were designed in the image of beautiful women—or impossibly beautiful characters. Rather, the gendering of AI VA is baked into the labor they perform, where they perform it, how, and for whom. The domestic sphere has long been thought of as a feminine stronghold, a site for women to control, beautify, and maintain. Even in a world where traditional gender stereotypes are challenged, gendering the new old housewives in association with the smart house relegates them to the digital interior. The devices are given meaning, utility, and a place in this world according to their domestic occupation as housewives.

In this chapter I described AI VA as the "new old housewives," a term that denotes the polysemy of the smart wife as a cultural agent and a mythic imaginary. The term "new old housewife" is descriptive of AI VA precisely because of its conflicting qualifiers; although these devices are technologically advanced, they connect to nostalgic, conservative, and old-fashioned narratives about women's roles in the household. The mythic housewife as a category of meaning and a domestic imaginary partitions the racial history of domestic work and sidesteps questions of class in exchange for a vernacular of ease and nostalgia. At the same time, it obscures the feminized work of raced and classed bodies that have historically supported the running of the household, albeit from a less visible vantage. Indeed, AI VA like Siri, Alexa, Cortana, and others are subject to a particular form of feminization, in which they are "demure, helpful and lack a threatening body."[98] At the same time, Siri and Alexa are asked to "perform stereotypical feminized labor," doing the type of care work the mythic housewife might have been expected to do at the height of the nuclear ideal.[99] AI VA are socially constructed actors given meaning through raced, heteropatriarchal domestic narratives that reinforce gendered divisions of labor and identity. This regressive and patriarchal framework "perpetuates superficial traits of femininity as the status quo."[100] Moreover, because AI VA are coded with a persona, a shadow of subjectivity, their inculcation as docile subjects in the domestic sphere may transfer to gendered bodies with which they coexist in the domestic space. Living in digitality requires the perpetual expansion of connected technologies, but it also necessitates the expansion of domestic ideologies. Biased AI VA set the discursive groundwork for categorizing and subsequently objectifying literally anyone who does not fit the dominant, racialized, heterosexual domestic ideal.

A related claim of this chapter is that the gendered persona of AI VA is deeply rooted in social construction of domesticity and domestic space. The role of the new old housewives is, to use Enoch's term, "a domestic occupation" because their utility and import is inextricably connected to historical narratives of the domestic space. In other words, AI VA's domestic occupation links the domestic work they perform to the spaces in which they perform it. The case studies show how AI VA are embedded in smart speakers, smart devices, and wearable technology that cook, clean, and sometimes even satisfy users sexually. Even if they are not overtly sexualized, they are sexed and gendered by their location within domestic technologies and the expectations that they will provide companionship as well as domestic work. The new old housewives, therefore, uphold a decades-old domestic sensibility that assumes that (white) women belong at home and at the express service of others. Their technological capacities, paired with approachable and nostalgic gender identities, make a spatio-rhetorical argument for living in digitality.

But the new old housewives have had enough. In response to the pernicious and unnecessary gendering of AI VA, scholars, activists, engineers, and others are demanding an alternative future. A report issued by the United Nations advocates for gender parity in tech development and through "gender-responsive" skills training. The report argues that "Siri's 'female' obsequiousness—and the servility expressed by so many other digital assistants projected as young women—provides a powerful illustration of gender biases coded into technology products, pervasive in the technology sector and apparent in digital skills education. This publication seeks to expose some of these biases and . . . begin closing a digital skills gender gap that is, in most parts of the world, wide and growing."[101] These interventions into education and the technology sector offer concrete steps toward a more just AI future.

Further, in response to entrenched gender bias in AI VA, commentators are calling for gender neutrality in digital assistants. The basic premise of these critiques is that customers should either have a choice in the perceived gender of their devices or that devices ought not to have a gender at all. Instead, activists, researchers, and practitioners advocate for creating gender-neutral characters (or AI/bots where gender is never revealed). Indeed, a gender-neutral assistant already exists in "Q." According to Q's creators, Q is a gender-neutral digital assistant solution that could later be embedded in most extant products using a gendered assistant.[102] Q was developed by a consortium of researchers, activists, and technology leaders advocating for a more just and equitable AI—but also a more just and equitable technology structure overall. Many of these contributors not only are experts in the field of AI and technology but represent individuals not typically represented in the field of technology industry and development.

Q's creators are on a quest to make Q's voice a "third option" in already-existing smart products. They have asked users to get involved in this mission by advocating for Q to Apple, Amazon, Google, and Microsoft.[103] By petitioning leaders in the digital assistant space, they hope to be able to implement Q's voice on the hundreds of thousands of Apple HomePods, Amazon Echos, and Google Homes already in use. Getting Q inside voice-assisted products is not only an exciting proposition but a strategic one. In terms of uptake, baking Q into these devices is strategic because they are among the most widely used. Petitioning Amazon, Google, Apple, and Microsoft to use Q could also help slingshot the problems with gendered AI into the public zeitgeist.

Q's creators argue that "technology companies often choose to gender technology believing it will make people more comfortable adopting it.

Unfortunately, this reinforces a binary perception of gender, and perpetuates stereotypes that many have fought hard to progress. As society continues to break down the gender binary, recognising those who neither identify as male nor female, the technology we create should follow. Q is an example of what we hope the future holds; a future of ideas, inclusion, positions and diverse representation in technology."[104] Q is defined in direct relationship to the new old housewives; Q is "like Siri or Alexa but neither male nor female." What makes Q different from Siri or Alexa is the pitch of the digital assistant's voice. Q's voice is pitched at 145–75 hertz,[105] a frequency band that is presumed to sound gender neutral.

Scholars of rhetoric and sound have noted the social and political importance of vocal sound. Voice is a key marker for identity and can be inflected with assumed racial, gendered, and other meanings. Amanda Nell Edgar argues that "the unique sound of an individual's voice often communicates politically coded information about the speaker, a process aided by cultural assumptions that characterize the voice as unconstructed and natural."[106] In humans (at least), vocal sound is influenced by "the body's composition and movement," which are themselves influenced by broader processes of social discipline.[107] Power influences vocal sound, vocal identity, and vocal privilege "from the outside in."[108] While voice assistants do not have the same bodily configurations as the humans they were designed after, their voices are still ideological in their reference to embodiment—they still call upon and reify social norms and hierarchies sonically. As such, focusing on the vocal register of AI VA—and the disciplinary ideologies they uphold—is a necessary intervention into the problem of gendered technology.

Q's contribution to the smart home discussion and to the conversation about gender in AI is significant. The introduction of a genderless device draws attention to the fact that devices are needlessly gendered in the first place. Offering an alternative clarifies the stakes of the debate and highlights the problems inherent in patriarchal systems. Moreover, Q's development is a positive step forward on inclusive AI politics—all necessary to build a more just smart home future. While projects like Q are important, they should be joined by affiliated movements for gender and technology justice.

We can move beyond voice to challenge the fundamental social, political, and economic structures that made gendering the digital assistant reasonable in the first place. Changing pitch may increase representation within gendered devices, but it doesn't actively work to undo the gendered, raced, and classed history of labor. Nor does it directly critique nostalgic, patriarchal domestic sensibilities. In her "Manifesto for the Broken Machine," Sarah Sharma guards against the idea that gendered devices can be rehabilitated incrementally.[109] This includes reworking assistive devices to be better representative of normative gender politics. For her, these assistive devices reify status quo politics, or what she calls "patriarchy's technological conditions of possibility." Instead, we ought to live as Broken Machines, the "radically inconvenient," a neo-cyborg that reimagines the relationship between power, society, and technology. As she writes, "broken machines do not operate under the logics of binary gender code and normative relations of care . . . they are not compatible within this larger system of new technologies. . . . Broken machines will not on insist on better representation alone."[110] Attempts to rehabilitate flawed AI VA may actually contribute to the project of "smart wifery" insofar as the device (and its AI persona) are asked to "'solve' deeply gendered (and . . . racially uneven) social problems."[111] Agitating toward AI voice gender neutrality must be paired with a structural critique of the material significations of gender, of which voice is only one. Examining the transfer of discursive residue between gender and the home space makes a spatio-rhetorical critique of the new old housewives particularly fruitful.

As such, I have argued that we must also consider the spatial occupation of gendered AI. We need to fundamentally upend the oppressive structures encoded into the gendering of virtual assistants by recognizing their positionality within the smart home. This means examining the extractive patterning of patriarchy in the development of these devices. But it also means revealing other invisible paradigms that influence gender, technology, and society: the invisibility and undervaluing of feminine and domestic labor, the inculcation of gendered stereotypes in novel domestic products, the re-creation of nostalgic and conservative domestic traditions in AI VA. To build a better virtual assistant, we need to rethink the smart home. This is the work of the next chapter.

4

What Smart Homes Can Be

In this book I argue that smart homes are the living manifestation of a digital imperative, one in which humans are expected to augment all parts of their everyday reality with platformed technology. Spatializing dominant discourses around domesticity and smartness, smart homes enact and reify a sociotechnical, determinist approach that justifies-in-advance the creation and adoption of new technologies in the domestic space. Once in place, digitality trades its initial foothold in the domestic sphere for a multiscalar expansion outward toward the public sphere. Although smart homes are frequently marketed as solutions for individual use, the case studies in this book evidence that the smart home operates at the level of the social. This means that even those who elect not to purchase, install, or calibrate smart home technology are likely to be affected by the worldview it constitutes.

The smart home serves as a nodal point for both technology and ideology. Individual smart devices in the home link to networks and existing infrastructures outside of it. Domestic ideals aid in the transition of computational logics to hegemonic social logics. Rather than living alongside or engaging with connected technology, digitality architects what at the outset appears to be a totalizing experience that becomes so encapsulating it is difficult to see beyond it. I have called this condition "living in digitality," a phrase that gestures toward the constant presence of smart, connected, or networked technology in space and in the public imagination. As I describe in the introduction of this book, living in digitality is a transformative experience stitching together social forces with technological capacity. Moreover, living in digitality is the immersive expression of platform logics governed by digital capture, capital, and control. Put simply, our current understanding of living in digitality is in service to the technological hegemons who manufacture both technologies and the discourses that justify their use.

This concluding chapter wrests the experience (and imaginary) of living in digitality from dominant platform logics and replaces it with a more inclusive cultural imaginary articulated through intelligences of care and community. Divesting from dominant narratives about the smart home, this chapter focuses instead on crafting a new language of intelligence that eschews profit over people. Mobilizing bell hooks's concept of "homeplace," the chapter celebrates the political nature of the home as a site of "liberatory struggle." Instead of defining the smart home in terms of accumulation, control, or optimization, I propose an alternate framework for the connected home: *the smart homeplace*.

The smart homeplace is an intentionally optimistic effort to reclaim the smart home as a site of liberation, recognition, care, and community. Rooted in the radical possibility of hope, this concept recognizes the power of collectives to reshape ongoing narratives about the value and utility of assistive technologies. From there, the concept of the smart homeplace provides a discursive platform to translate our new shared imaginary around smartness into concrete practices that resist the oppressive forces of technoliberalism. Concurrently, the smart homeplace is an invitation to investigate and prioritize intelligences beyond computation. Bodily intelligences, collective intelligences, and underrepresented and marginalized intelligences provide an expansive framework to reimagine the home of the future. Spatializing these intelligences into the smart home can productively interfere with living in digitality as a hegemonic force. Thus, this chapter offers the smart homeplace as a heuristic for reevaluating frameworks of value relative to digitally connected technologies.

Extending the smart homeplace concept into the future, this chapter incorporates demands from technology and domesticity scholars and activists to reimagine the smart home as a site for radical change. As a cultural imaginary, the smart homeplace provides touchpoints for building a smarter, more inclusive domesticity. Some of these touchpoints are practical, and others are utopian. By meshing these two orientations within the framework of a feminist ethics of care we can begin to recuperate a smart domesticity for the collective rather than the individual.

This chapter proceeds in three parts. First, I describe how smart homes configure smart domesticity in the image of a centuries-old domestic ideal. This domestic ideal is so durable that it limits imagining future possibilities for the smart home. Outlining these constraints is the first step to crafting a more just, progressive, and inclusive smart home future. In the second section, I briefly return to key points from case studies throughout the book, explaining the stakes in letting the current paradigm of living in digitality to continue unencumbered. The third section examines possible smart home futures, including wholesale rejection of the smart home paradigm. Finding these solutions necessary but insufficient, the final section advocates for a smart homeplace, in which domestic technologies are reconfigured in the service of a new domesticity that foregrounds community and care over profit and technoliberalism. Let us begin by examining two forces organizing possible smart home futures: the durable domestic ideal and platform capitalism.

LIVING IN DIGITALITY AS A DOMESTIC IDEAL

Although platform capitalism is a powerful shaper of the modern smart home, it is not the only force ushering in a technoliberal smart home future. The flexibility and plasticity of the domestic ideal have also helped to organize the smart home future, domesticating digitality toward computational ends. As I have argued throughout this text, the domestic ideal is a durable configuration of meanings about home, belonging, autonomy, social reproduction, and consumption. These meanings manifest into spatial arrangements that reify traditional domestic discourses into physical places where people live, work, raise families, experience pleasure, and more. While ideal domesticity shifts to incorporate cultural changes, including the emergence of new technology, it attends to the needs, desires, and expectations of the dominant class. Domesticity is therefore durable in two ways: first, as a discourse, domesticity figures how people should live according to a set of vague but cognizable ideological standards. These standards support a domestic ideal configured around white, heteropatriarchal privilege, which purveyors of these fantasies sell back to everyone else in both overt and covert ways. Second, as a cultural ideal, dominant configurations of domesticity become a technology all their own. Domesticity mediates numerous forces and institutions both within the home and outside of it, drawing together social norms, cultural expectations, economic systems, national imaginaries, and more. This explains why emerging technologies that could radically reorient the domestic sphere in alignment with justice initiatives are repeatedly domesticated by forces like heteronormativity and platform capitalism.

Key to this domestic ideal is the persistent imaginative and literal policing of the domestic interior, a hermetically sealed and protected retreat from the world. The idealized domestic interior presumes in advance the individualized autonomy of a singular family unit, distinct and separate from both the public sphere and other family units. Although the domestic ideal is not always (or even often) reflected in the spatial reality of housing or in the everyday lived experiences of people, it continues to shape the collective imagination of the modern smart home. As Heckman argues, "home, in the material sense, for the American middle class is typically imagined as the single-family, suburban house, but has increasingly come to include apartments, condominiums, townhouses, trailers and any housing option that affords a somewhat autonomous living space organized around the head of the household or the family unit."[1] The domestic ideal manifests in various spatialized formats, stitching these cultural figments into physical space whether or not that space resembles the American Dream home. Frequently, such spatial configurations affirm the implicit norms held within the domestic ideal. These implicit norms speak to questions of status, ability, class, and more. As evidenced by the proliferation of "smart" facsimiles of midcentury middle-class homes, this domestic ideal continues to influence the smart home imaginaries of today.

The way that most people experience it, the smart home affirms normative configurations of domesticity. Markers of American domesticity, such as self-sufficiency, privacy, personal choice, and more, are baked into larger nationalist narratives of home in the United States. Scholars have argued that the domestic imaginary is inextricably linked to nationalist imaginaries. Amy Kaplan, for instance, has argued that "domestic has a double meaning that not only links the familial household to the nation but also imagines both in opposition to everything outside the geographic and conceptual border of the home."[2] Whatever is inside the home is cherished, intimate, and protected. Whatever is outside the home is understood as foreign, other, and (potentially) dangerous. In this configuration of domesticity, the domestic ideal offers affective cover to draw divisions between self and other, between family unit and the community. The family unit comes out on top—both in the private sphere of the home and in the public sphere more generally. Lauren Berlant highlights the connection between the domestic ideal and citizenship through the figure of the "intimate public sphere," noting that the family mediates a utopian promise of citizenship that belies publicity and affirms the intimacy of "nostalgia-based fantasy."[3] This fantasy foregrounds the fictive family, around which nationalist identity coheres. Domestic fictions animate the home's domestic interior even as the domestic ideal shapes the experience of spaces beyond it—including the public sphere.

Smart home technologies further atomize familial relations in the domestic sphere, prescribing and normalizing hierarchical relationships within the home. As Murray Goulden suggests, logistical structures in smart home accounts produce "the platform family," or a rearticulation of the domestic unit through the lens of platform capitalism. The platform family foregrounds and reifies the normative figure of "co-located nuclear families."[4] This type of domestic structure, while flexible within certain configurations of platform capitalism, ultimately draws upon and intensifies heteropatriarchal logics of both family and home.[5] Home is made meaningful not as a space of care and concern but as a site for consumption and extraction. Platforms are implicated in this consumption in several ways—as sites where users can purchase goods, but also as infrastructures that collect, prepare, and monetize user data. The platform family is spatial, discursive, and recursive: it implicates the domestic unit within the platform economy by organizing the familial unit in its image. The siren call of smart domestic technologies is durable but incomplete. Rhetorically, mobilizing the figure of "home" strategically supports technoliberalist agendas by literally domesticating them. The impact is especially pernicious, because it is as expansive and mutative as platform capitalism is. As Goulden notes, "by inserting itself within domestic practices, by capturing material and symbolic elements of it, the platform family seeks to supplant social forms which currently lie outside their valorization systems. Accordingly, consumerism is awarded primary position in this new form of doing family."[6] Smart technologies play on this fantasy by borrowing domestic fictions of the past and revisioning them through digitality.

Instead of changing what domesticity means, or who domesticity is for, in meaningful ways, the modern smart home (re)makes the domestic space more valuable for platform capitalists, who use the space to learn about the habits of users, track their movement throughout space, and sell that information back to them (or to others).

Consider, for instance, the persistent problem of domestic labor. While smart homes purport to make life more convenient and efficient for householders, they do not fundamentally augment domestic demands. Smart technologies that purport to make life more efficient and leisurely may do so by interlinking larger infrastructures of platform labor or outsourcing them to coded objects. It is not that the smart home has fundamentally reconfigured labor for the benefit of workers—it's just reworked it through the figure of the platform. For instance, in chapters 1 and 3, I describe the development of a new workplace in "the drop zone," and a new set of domestic workers in virtual assistants. With the assistance of smart technologies, algorithmic logics, and massive corporate logistics infrastructures, the basic processes of selecting foods to eat, buying them, transporting them, and then cooking them has been stitched into the logic of the platform. There, lean platform economies outsource the labor to increasingly precarious workers. At the same time, this process relies on the figure of a virtual assistant, a disembodied other given a gender to make these platform transitions appear banal.

In fact, smart technologies can intensify domestic expectations while making the *work* of the smart home less visible—a lesson learned in the Drop Zone of Brookfield Residential model homes in chapter 1. The invisibility of the smart home is a significant aspect of the smart home sell. The Brookfield Residential Smart Home visited in this chapter is not manifestly experimental in aesthetics or function—it looks, feels, and acts much like the bourgeois domestic ideal that emerged in the nineteenth century and has been refined to the present moment. Whereas other "homes of the future" worked to uproot architectural and cultural norms about the meaning and function of the home, the Brookfield Residential site reinvested in traditional understandings of "home." From a spatial rhetorical perspective, it is manifestly clear that the Brookfield Residential Smart Home is built, marketed, and arranged to be the physical manifestation of a middle- to upper-class American domestic imaginary.

As the case studies in this book show, smart homes traffic in spatial rhetorics that purport to provide ease and safety from the outside world but also increase connection to it. The smart technologies in the space do double work. They segment the presumed nuclear family unit from an unknown, presumably dangerous outside world. At the same time, smart technologies network the home within digital infrastructures, public utilities, and platform technologies. Traditional/transitional smart houses secure the home from the rest of the community, as the experience of connection is refigured through the infrastructure of the platform. In experimental smart homes, the radical pull of alternate forms of domesticity is diminished by forces of austerity and technological solutionism. Retrofit smart homes boast similar effects but may express them incrementally as the platforms are introduced to a house one device at a time.

Smart homes are by definition connected homes. Here, connectivity—both technical and social—is mediated within platform logics, recast in controlled architectures of digitality. Digitality is figured as a perpetual and obligatory status even when it results in alienation of the self and the other. The platform's appetite has no bounds, as evidenced by the proliferation of smart technology (some of which appears to have no purpose at all.) Rather, the function of digitality is to incorporate all aspects of living under the sign of the platform. The process of incorporation is necessary to maintain platform capitalism. As Nick Srnicek argues, "not all—and not even most—of our social interactions are co-opted into a system of profit generation. In fact one of the reasons why companies must compete to build platforms is that most of our social interactions do *not* enter into a valorization process."[7] The home is culturally valorized because it is a lucrative site of data extraction for expansive platforms in hot competition to collect, quantify, and monopolize user data. The smartest elements of the home outfit the space with technologies that connect the individual home to platform capital, including Amazon, Instacart, and other subsidiaries. These technologies relay domestic needs—like acquiring groceries and other domestic goods or feeling a sense of connection and accomplishment—through the machination of the platform.

As seen in chapter 2, even those experimental smart homes that push against normative standards of home are disciplined by internal and external demands on the smart home. The Duke University Home Depot Smart Home is radical insofar as it reconfigures private living space for public problem solving. The smart home is smart not necessarily because of its technology (which becomes obsolete at a dizzying pace) but because it draws together nonfamilial individuals to live and work sustainably. Still, the moniker of "smart" proves to be a useful fundraising framework as well as a way to incorporate outside stakeholders (including greenwashing corporations) in the work. This is important for universities facing budgetary challenges, even wealthy ones, which must prove their utility to several, often conflicting audiences. Although the sustainability work of the Duke University Home Depot Smart Home is existentially valuable for a world facing the ticking time bomb of climate change, its smart mission must be made palatable (and productive) to a skeptical world run through capitalist logics.

Imagine! Colorado Smart Homes also uses the framework of the smart home to attract attention, resources, and support from outside stakeholders. The Charles Family Smart Home, visited in chapter 2, troubles the digital domestic ideal by creating safe, accessible, and inclusive living and working spaces for individuals with intellectual and developmental disabilities. The smart home is organized according to the needs of the individuals within it, rather than conforming to a nostalgic configuration of single-family domestic life. Smart technologies are part of the solution insofar as they can respond to the myriad ways that homes are disabling rather than enabling for diverse bodies.[8] Yet, like the Duke University Home Depot Smart Home, the smart technologies needed by residents are expensive and risk becoming obsolete before funding is made available for them. For this reason, the Charles Family Smart Home uses cheaper, more available, off-the-shelf technologies, including Apple iPhones and Amazon Echo devices. When deployed strategically in support of inclusive domestic practices, these devices can supplement highly specialized medical technologies—or serve as a holdover until those technologies are available and paid for. The use of these devices presents as a hacking ethos, which as a practice is laudable but is troubling insofar as it is ultimately necessitated by insufficient public and private funding for residents.

REMAKING THE DOMESTIC IDEAL BY RETHINKING THE SMART HOME

Outfitting our homes with smart technologies is not a neutral proposition. When we buy a smart device, plug an existing appliance into a smart outlet, or engage with Siri, Alexa, Google Home, or another virtual assistant, we are interfacing with both the domestic ideal and a larger platform, which has a politics alongside its institutional policies, norms, and rules. For many of these technology hegemons, the North Star for smart home development has been dominance in the data economy—profit and expansion through data extraction and monetization. This approach to digitality manifests in smart technologies that "further the aims of data collection and social control."[9] Purchasing and using smart devices in our homes is therefore a tacit acceptance of the consequences of living in digitality. As Sadowski observes, "for every smart self-tracker, there are data brokers and bosses who see new tools to exploit people. For every smart home appliance, there are manufactures and insurers that see new opportunities to extract value. For every smart city solution, there are police departments and platforms that see new methods to exercise authority."[10] And although we may consent to sharing some of our daily life with these devices, the positive and pernicious effects may not be readily apparent.

In this book I engage in embedded fieldwork and subsequent rhetorical criticism of smart homes as a narrative for living and as a set of spaces to physically occupy. Attending to the spatio-rhetorical dimensions of the smart home has drawn me to characterize living in the smart home as a cultural condition, rather than an individual experience.[11] To that end I introduced the concept of *living in digitality* as a lens to identify and trace the sociocultural effects of the smart home. Often smart homes promise a solution to the trials of the modern domestic condition. By using computational technologies and computational principles of ordering the world, smart homes promise an elevated experience of home. Rhetorically framing the smart home toward individual benefit is particularly strategic for those who make and profit from the smart home. But individualized frameworks of value tend to miss the more structural effects of digitality as a social force. In chapter 1, for instance, I demonstrate the extent to which smart technology normalizes the obscuring of platform workers while simultaneously subjecting them to intense monitoring. Chapter 3 shows how virtual assistants like Siri and Alexa's encoded personas reify gendered stereotypes, opening the door to gender violence against their human counterparts. Chapter 2 describes how socially conscious experimental smart homes have difficulty slipping the grip of digitality's optimizing principles. Even as these living laboratories work to undermine hierarchical structures, their attempts are shaped by social, political, and economic forces of digitality beyond the smart home.

In analyzing the phenomenon of living in digitality, I make three other, related claims that lay the groundwork for this chapter. First, I suggest that the smart home's technological imperative manifests spatially, by organizing domestic interiors in accordance with the logics of the digitality. Indeed, if, as many communication theorists claim, symbol use can shape reality, the spatialization of digitality in the smart home is a concrete case of how this might work. The case studies in chapters 1 and 2 demonstrate the extent to which smart technologies "transduce space" in accordance with the demands of the platform.[12] Chapter 2 highlights more significant changes to the makeup of the domestic interior made possible by smart technologies. That these spatial alterations can be subtle or radical evidences the flexibility of digitality and its many entry points into domestic spaces. For our purposes, spatialization extends the horizon of action for activists, users, academics, and politicians interested in interfering with hierarchical and exclusionary models of digitality. Moreover, the spatialization of digitality in the smart home presents a practical example of how digitality might manifest in other spaces beyond the domestic sphere. Thus, while this chapter imagines alternate frameworks for connected living in the domestic sphere, stopping our interference at the smart home leaves the roots of digitality very much intact.

Second, and relatedly, I argue that living in digitality is a generalizable condition connecting the home to spaces beyond it. This means that the smart home is networked not only by discrete technologies, but by cultural, social, and economic infrastructures that support its daily operation. These structures also provide frameworks of value that justify the smart home concept and produce the ideal subject to live within it. This conclusion may appear somewhat paradoxical, as much of the public discourse on smart homes is articulated at the level of the individual user. Yet smart homes commodify domesticity as social force by bringing the home into persistent, consistent encounters with platforms. Smart devices are an umbilical cord between massive technology corporations and intimate parts of our lives. Living in digitality presently supports a larger system of platform capitalism, the most recent iteration of capitalism that structures the economy around the extraction, processing, and monetization of data. Smart homes are crucial battlegrounds in the so-called platform war, fought by technology hegemons such as Amazon, Google, Apple, and others. Combined, these cultural conditions suggest that these platforms are likely to continue unabated in their quest for smart home domination, absent a significant, persistent, and concerted effort to challenge the structures of platform capitalism.

The third and related claim made by *Threshold* is that despite efforts to brand them as such, smart homes' definition of "smartness" is predicated on a narrow and exclusionary understanding of intelligence. Smartness here stitches together intellect with cognitive capacity, and cognitive capacity with rationality and objectivity. This definition draws on legacies of ableism, eugenics, and white supremacy while also rejecting other forms of non-normative intelligence. In the case of the smart home, smartness as concept is profoundly computational in nature, and it is a type of computation defined in the image of capitalist accumulation. When living in digitality, smartness is dedicated to automation, incorporation, and quantification. Discursively and ideologically smartness serves as narrative structure determining the utility and value of all human and nonhuman practices. Practically, this discursive framework manifests in smart home products that are narrow in scope, limited in capacity, and set up to affirm or intensify the status quo—not change it.

As a result, although smart homes are marketed as forward-thinking and futuristic, most smart homes direct smart technologies to solve very limited problems—or create problems to solve them. Jathan Sadowski has observed, "What do we get from all this innovation? We mostly end up with an overabundance of solutions looking for problems."[13] For instance, the Brookfield Residential Smart Home has inaugurated a new set of tools and domestic spaces to solve the supposed problem of "porch pirates"—a novel condition exacerbated by the proliferation of online shopping and platform-managed delivery. The smart appliances equipped with Siri and Alexa to cook and clean don't fundamentally challenge domestic sensibilities or expectations; they simply offload them onto a new digital agent with the same domestic occupation. Smart homes are largely breeding grounds for capitalist accumulation: of stuff and experiences and services for smart home users; of user habits, preferences, relationships, and desire for platform capitalists.

The nature of the platform is to serve as an intermediary, further connecting us to other systems, services, and infrastructures, which are often hidden from our view.[14] Scholars have noted how technoliberalism organizes our everyday lived reality through an intensification of digitality in all spheres of life. In their "Five Theses on Technoliberalism and the Networked Public Sphere," for instance, Damien Smith Pfister and Misti Yang argue that technoliberalism serves as a "governing rationality in cultures where digital computation technology suffuses everyday life."[15] This governing rationality operates at the level of the marketplace and the state, but also in more banal spaces like the home. In the home, technoliberalism appears as a shiny, understated imperative to use technology to upgrade one's life to the "good life."

Technoliberalism supplants collective decision making, undermines democratic practice, and rejects community action. Rather than supporting democratic deliberation, technoliberalists prefer instead to "design complex systems to redirect behavior" in ways that prop up platform capitalism.[16] Smart homes are an exemplar of these complex systems, because they reprogram the functioning of the home by locating the platform inside of it. Smart homes redirect human behavior through the mythic utopia of ubiquitous computing, wherein the platform becomes the dominant structure of life lived at home (and beyond it). When users route basic household tasks such as cooking, cleaning, shopping, parenting, and more through smart devices, they are lending themselves—their bodies, their attention, their capital—to the significant influence of the platform economy. And although technoliberalism gestures toward the separation of politics from technology, technoliberalism incentivizes an intensely hierarchical system of winners and losers, in which the most fundamental components of human life are commodified, tracked, and controlled by computational logics.[17]

The smart home is an exemplar of digitality's capacity to augment human behavior in ways that benefit platform capitalists and harm communities. Rather than serving as a refuge and respite from the world, the smart home instead becomes part of the platform infrastructure. So, too, the humans inside of it are subject to the mechanized pull of the platform. The home becomes a marketplace where householders buy consumable goods, manage invisible labor-others, and ambiently produce valuable data to be extracted, prepared, and monetized by platform capitalists. To be sure, for a few privileged subjects, smart homes can provide some ease, convenience, and leisure. However, platform capitalism (and technoliberalism writ large) is an inherently extractive, violent system. For the vast majority of digital subjects, any smart home benefits are short-lived at best. Moreover, the freedom offered by such technologies—sold to us by technology hegemons—comes at the cost of community and of collective care.

Here are the risks: we are trading social justice for momentary convenience. Our ease is paid for by people who live in poverty, who work in increasingly unsafe conditions, who labor contingently, without healthcare, without days off, without opportunities for collective bargaining. Our supposed security comes at the cost of the broader collective, because dominant smart home narratives assume that our community members are the primary instigators of violence. In reality the violent structures and systems that activate divisions, mistrust, and hierarchy are responsible for a collective lack of safety. In other words, convenience, ease, and security are abstract ideas used to sell smart homes, rather than concrete consequences of their usage. For most of us, these ideas currently or will operate only in the realm of the abstract. Ubiquitous computing has not freed us from bounds of gatekeepers, hegemons, and technocrats. Rather, it has deeply embedded us within their grasp. In sum, at present, living in digitality prioritizes capital and data accumulation over other, more inclusive domestic values such as care and well-being; in so doing the smart home subverts the most radical aspects of domesticity for profit. But this is not the only way to imagine and live in domesticity. The smart home as it exists now need not exist in perpetuity. Rather, we can intentionally construct new imaginaries for the smart home.

Remaking the smart home is a multifaceted process, which oscillates between small pragmatic steps and large structural change. Lauren Berlant supplies the concept of interference as a way to reconfigure unsuitable frameworks around nationality and citizenship, but the practice can also be applied easily to the related concept of living in digitality.[18] Interfering with normative conceptions of smart domesticity asks us to unsettle constructed but durable narratives around domesticity and smartness. As Berlant writes, "this kind of writing does not aspire to merely comment on or contextualize its object."[19] Rather, through a process of "radical recontextualization" of smart homes, critics can make dominant smart home narratives and "the political culture they operate in look unfamiliar and uninevitable."[20] Both steps—making an object strange and rejecting apparent sociotechnical inevitabilities—are crucial steps to constructing a new, progressive cultural imaginary. The process of interfering is inherently active and provocative, rejecting dominant reality structures by imagining an alternate pathway forward. For those who study the ways that words and other forms of discourse gain meaning, naming is a politically charged action that demonstrates one's worldview, values, and assumptions. Further, seeking ways to reclaim (or recontextualize) smart homes outside of a technoliberal context provides room to dispute the inevitability of a particular technological future. Interfering is, by its nature, hostile to the supposed teleological condition of platform capitalism. As a process it demonstrates an interest in obstructing, stopping, impeding. But in making the smart home strange, there is also an "after," a something else that is created to fill the gap of the unsuitable thing.

As a part of the process of interfering, I propose we use the framework "living in digitality" to describe the social and technological system in which digital technologies increasingly mediate all aspects of daily life. In describing the features of living in digitality, I put a premium on showing the ways that digitality is *agential*; that is, it shapes conditions of being and acting in the world. The smart home enacts the features of digitality in domestic space, concretizing computational values, and embedding them into the walls. And yet digitality as a framework for living is not limited to the smart home setting. Rather, digitality is a much more expansive force that shapes spaces and social relations at a much broader level. Said a different way, living in digitality is characterized by its environmentality. It is an extensive and immersive condition in which computational technologies engender a particular worldview, and that worldview is spatialized in everyday settings.

The insidious, multiscalar nature of digitality implicates systems and collectivities as much as it does individuals. Said another way, living in digitality affects individuals who elect to engage with smart home technologies, and it also affects individuals who do not. This fact presents challenges for remaking the smart home. On the one hand, there are practical actions that readers can engage in today to reject the biased and totalizing frameworks of digitality in the smart home. On the other hand, digitality as a structuring apparatus that manifests in smart homes but also in other social settings, necessitating a similarly structural intervention.

Undoing the damage of the technoliberal smart home might begin first by rejecting the language used to describe the smart home. Sadowski, for instance, advocates against "rehabilitat[ing] and appropriat[ing] the label" of smart home. Instead, he suggests we should "disavow *smart* as a paradigm."[21] The strategy of disavowal proposed by Sadowski includes "unmaking." Unmaking forwards a neo-Luddite technopolitics in which users "downgrade" their homes, removing smart technologies. In practice, smart home unmaking might include abstaining from the devices, services, and other related processes that make a smart home smart. These individuals may choose simply not to purchase said devices, abstain from the use of any devices with Wi-Fi connectivity, or decide against the use of the "smart" aspects of the devices. Users could choose to not connect their Spotify account to their refrigerator, never conjure up Siri or Alexa (or Cortana or Erica), and reject persistent prompts to connect this or that device to their internet network.

A more radical version of unmaking might be a total retreat from the smart home concept and, subsequently, smart home space. This strategy has the benefit of being radical, and it would send a very clear signal to technology hegemons that individual consumers reject the smart home as a project. Historically, rejection through retreat has been a productive protest mechanism, particularly if done en masse. However, while a tempting solution to the smart home problem, the conditions of living in digitality threaten to make retreat from the smart home an option solely for already-privileged subjects.

Frankly, it already is. Consider, for instance, boutique services that purport to protect users' privacy—for a fee. In a 2014 op-ed for *The Guardian*, writer Arwa Mahdawi explains that "the digital divide is becoming a privacy divide." Mahdawi notes that a new industry has formed around growing fears about privacy and tracking in the digital age. "Solutions" offered by this new industry include a high-tech phone equipped with encryption features as standard, and reputation management services that undertake the laborious process of removing one's personal information from the internet. Other, more low-cost privacy tools, like ad blockers, VPNs, or privacy-sensitive search engines, require knowledge and skill to find and incorporate. As Mahdawi argues, "These tactics tend to be impractical, complicated or require a level of tech literacy most people simply don't have. If you are using a virtual private network or seeking out more secure alternatives to Google, it's more than likely that you have a certain level of education and come from a higher socioeconomic background." In other words, while tools exist to manage a user's relationship to the pernicious elements of living in digitality, they are not equally available or distributed across populations.

Living in digitality has adapted to incorporate retreat as a feature rather than a bug of an ever-expanding techno-politics. In her book *How to Do Nothing: Resisting the Attention Economy*, artist and writer Jenny Odell evaluates these retreats, linking "digital detox summer camps" to hippie communes of the 1960s. While both practices affirm a radical politics that is productive in theory, the overall effect of retreat can leave violent systems and structures in place, especially for those who cannot retreat. Says Odell, when faced with the overwhelming experience of capital-technology, "We might be tempted to conclude that the answer is to turn our backs to the world, temporarily or for good. But this response would be shortsighted. All too often, things like digital detox retreats are marketed as a kind of 'life hack' for increasing productivity upon our return to work. And the impulse to say goodbye to it all, *permanently*, doesn't just neglect our responsibility to the world that we live in; it is largely unfeasible, and for good reason."[22] While there are differences in scale and intensity, choosing not to live in a smart home or engage with smart homes might be an extended form of a digital detox retreat. Like other digital detox retreats, it leaves in place the social, political, and economic forces that produced the smart home in the first place. Moreover, "life hacking" the smart home by retreating from it manifests digitality's optimizing narrative by providing a temporary but palatable "solution." While optimization often looks toward short-term solutionism in accordance with capitalist narratives of value, the trouble with simply abstaining from the smart home is that the technical and social networks that give rise to it do not simply end. Rather, as seen throughout this text, living in digitality is a structural principle that augments life inside and outside the domestic exterior in a profoundly general way. Given the insidious and expansive nature of digitality, a smart home detox or retreat is an insufficient solution. In fact, these attempts to retreat can be used to reincorporate users into a new technological scheme with a slightly different user interface (e.g., the historical transition from "home of the future" to the "smart home" or its future equivalents).

Let us consider some practical realities of living in digitality. It is now becoming difficult to acquire uncoded, not-smart versions of technical devices. Many appliances, devices, and gadgets now come with smart technology standard. At the same time, more and more smart devices are being invented and sold (or retrofitted) to extend digitality's reach. Currently, digital connectivity is generally seen as a premium benefit. Smart devices can therefore demand a higher price point than their non-smart counterparts. However, this might not always be the case. In the very near future, smart devices might be the cheaper option based on an already-existing subsidization model, in which platform capitalists sell products at a cheaper price in exchange for insight into users' lives. Many smart devices are not profitable in their own right. Rather, they are profitable because of the insight and data they provide to parent corporations. It is well known, for instance, that Amazon subsidizes its smart devices, allegedly below production cost.[23] This process is standard in a platform economy, where platforms might cover a loss on one part of their platform in exchange for another benefit.[24] It is not difficult to imagine a future where smart devices are the most cost-effective options for a consumer, who tacitly agrees to exchange their data for a lower-cost product. If this were the case, privacy would be the premium feature. Abstention would be a luxury offered only to those who could afford it.

While there may be a choice not to engage with smart technologies, some groups will have more choice than others over the matter. Luxury tech-detox wellness retreats point to (temporary) sanctuary from digitality as a privilege afforded to those who can "disconnect," rest, and recharge. Others, including the widening portion of contingent workers in the gig economy, remain tethered to devices that track, quantify, and order their working (and nonworking) hours. The human alternative to digital optimization may also become a luxury good. As more customer service roles are outsourced to bots, curated, concierge human interaction is increasingly becoming a premium feature for valuable digital subjects. At the same time, workers who previously performed those roles are seeing their hours, wages, and benefits slashed. The smart home already affirms and serves valued digital subjects, in part by affirming their status relative to others living in digitality. Given digitality's expansive capacities, this hierarchy does not end at the smart home.

As we work toward the process of remaking the smart home, we must be vigilant about uprooting smart technologies that currently serve communities on the margins. Outright rejection of smart technologies ignores the fact that many technologies perform distinct and measurable social good for specific groups of people. For instance, smart technologies in the home can support aging and disabled folks in leading independent lives with dignity. Voice-activated devices can be a useful tool for individuals with "a variety of disabilities, enabling them to control devices through speech rather than through a screen interface or touch."[25] Smart technologies that track health or monitor a user's personal environment can be especially useful for individuals with health concerns—and their families. Smart technologies can also be resource-effective tools to supplement medical staff and services that are not accessible because of distance or other reasons. Until major structural changes in the medical and care economy are made, and houses are built to enable rather than disable individuals, a wholesale rejection of smart technologies could leave these individuals out in the cold. Instead, smart technologies must be designed with these individuals in mind.

While smart technology currently has a corporatist politics, it is not guaranteed that technoliberalism is its ultimate expression. As Pfister and Yang remind us, "technoliberalism may be the default public philosophy of digital culture, but it is neither natural nor inevitable. . . . Technoliberalism was built, layer of code by layer of code, and it can be rebuilt with a more democratic and deliberative ethos."[26] Sadowski similarly advocates for democratizing innovation and redistributing ownership of user data, reimagining technological innovation as a collective, not-for-profit endeavor in which the proletariat engages in "seizing the means of innovation."[27] Rebuilding the smart home necessitates digging into the social and technical frameworks that not only produced that code but sanctified it as the one true path toward domestic enlightenment. This process is both rehabilitative and radical in the short and long terms. It looks toward the past to see how we got to this place and looks forward assuming there are alternate possibilities for both technology and domesticity. Rejecting the corporatist foundations of the smart home does not mean rejecting the framework of technology in the home writ large. It means reclaiming it from the corporate hegemons who currently control its development, manufacturing, and usage.

WHY CARE COMES FIRST: REMAKING SMART DOMESTICITY ANEW

Smart homes are defined primarily through their technological capacity. Homes that are connected to the Internet of Things, or the internet more generally, can claim the designation of the smart home. As a result, those who manufacture these technologies have undue influence on the meanings and types of experience provided by the smart home. We can remake the future of the smart home by drawing upon critical frameworks that challenge not only current expressions of smart technologies but also the structural conditions from which they came. In particular, feminist approaches to domesticity serve as a useful scaffolding for a more just smart home future.

The goal here is not simple or modest reform of either digitality or hegemonic conceptions of domesticity. Rather, the goal is to meaningfully challenge the narratives of space that insinuate the smart home within technocratic social and political structures. A feminist spatio-rhetorical approach is uniquely situated to inform this project because it highlights the possibility of remaking and renegotiating meaning of already-entrenched ideas, perspectives, and objects. Spatial rhetorics are an especially appropriate framework for developing more-just domestic imaginaries because, as a practice and lens, it highlights how spaces can be made and remade in new light. As Jessica Enoch observes, "such spatio-rhetorical attention underscores not just the flexibility of space but also the imaginative possibilities for spatial creation so that access, inhabitation, and communication are possible in ways they may not have been before."[28] Imagining possibilities includes finding new ways to enact change in the smart home.

Since the home has been historically designated as the realm of the feminine, feminists have persistently located the domestic sphere as a site of social relations, social (re)production, and labor—and resistance. They have argued steadfastly that it is a site of politics worthy of public attention, both by critiquing hegemonic forms of domesticity and by attempting to remake domesticity in a feminist light. We can borrow some of their recuperative, resistive frameworks to imagine the smart home anew. Once organized, feminist approaches to smart domesticity can figure the experience of living in digitality away from corporate, technocratic forms of life. Instead, new domesticity reimagines the technological as a site for challenging white supremacist heteropatriarchal domestic ideals. Scholars and activists have noted that housing is a strategic and provocative entry point for structural change. Progressive housing movements "are united, in their difference, by their effort to use housing *as a gateway* to challenge wider structural forms of violence, including patriarchy, racism, class exploitation, and, of course, deprivation of shelter."[29] Our smart home future must therefore be an intentional, intersectional, revolutionary one.

Critical and radical feminist theorizing around the roles of domesticity and belonging is especially useful to the project of remaking smart homes. Feminist frameworks of domesticity can demonstrate how to remake space through an ethic of care and collaboration, one which recognizes the inner workings of home as having innate value before platforms decide it is so. Feminist theorist Valerie Padilla Carroll has examined domesticity in many of its dominant forms. Rather than seeing domesticity as a lost cause, she has sought ways to reenvision it. Padilla Carrol draws on feminist thinkers, activists, and writers who seek recalibrated domesticity to challenge heteropatriarchy rather than contribute to it. Her research suggests we find affinity in feminist, environmentalist thinkers such as Vandana Shiva, Sharon Astyk, Shannon Hayes, and Barbara Kingsolver. These figures "reject, or at least evade, capitalism and consumerism in favor of frugal and local living."[30] This radical project locates the home as a central site for reimagining a new politics. Domesticity is understood as scaffolding for this process; it is wrenched from hegemonic traditions of home and renewed with feminist values in mind. This "new domesticity" is figured as an anticapitalist, procommunitarian ethos that reshapes modernity beyond capitalism's vertiginous reach.

Feminist approaches to domesticity have manifested in "new domesticity." Padilla Carroll notes that new domesticity has radical capacity as a "potent revolutionary activism." [31] As I understand it, this "new domesticity" is a patently collectivist and antihierarchical figuration of the domestic sphere. It takes advantage of the home as the interface between the social, political, and economic spheres. New domesticity, Padilla Carroll writes, "moves beyond the self to the political, emerging as a radical, eco, and feminist housewifery that grew out of and in response to the US Women's Liberation Movement. . . . New domesticity promotes its altermodernity project: an anticapitalist environmental activism that embraces a global vision of sustainability based on reciprocity and care."[32] New domesticity is antithetical (even openly hostile) to capitalism and rugged individualism. Indeed, reciprocity and care underpin agency and empowerment of people in the home and beyond it.

New domesticity figures the home as a site of shared struggle against the violence borne by white, bourgeois, heteropatriarchal domesticity. Home here is both a symbolic fixture and a process of remaking. As a symbol of resistance, home is a way to reclaim space, time, and relation for the collective good. As a process of remaking, it provides the social and spatial processes to care for self, others, and environment. Whereas the domestic ideal of the nineteenth century helped solidify the discursive and spatial bounds of public and private, new domesticity focuses instead on sustainability through community.

New domesticity is an alternative to "global capitalism's consumer and extractive economy," where "people's empowerment comes from valuing the affective labor of the home, they find security in the community rather than the individual, and they develop economics of reciprocity rather than profit."[33] New domesticity is a radical proposition insofar as it rejects many of the durable principles of domesticity that have endured for centuries. It is a social project because it asks people to recognize the importance of the home and the people within it rather than finding ways to outsource it. Community grounds the experience of home, and shared experiences are supported through multiple forms of domestic labor.

Central to this effort is recognizing the importance of domesticity as a structuring apparatus for the experience of home and spaces beyond it. As Padilla Carrol writes, "recuperative feminist scholarship, often focusing on capitalist economies, situates domesticity as central to understanding gender inequality. Rather than perceiving domestic and reproductive labor of the home as economically unproductive, they argue for the vitality of such work because it reproduces and maintains the workforce."[34] The domestic sphere is important not only because it provides shelter but because it is a site of social and spatial reproduction. What happens inside the home influences what happens outside of it. The home is therefore a central nodal point within larger social, political, and economic structures. Engaging within it is to necessarily engage with the structural conditions that make smart homes damaging to the collective ethos of humanity.

New domesticity is one waypoint in reworking domesticity for the benefit of the majority. In building a better smart home, we will have to participate in this domestic labor together. That means traditional domestic tasks like caring for the home and for the people, animals, and other living beings within it. But it will also mean creative capacity building for rebuilding more inclusive, egalitarian, and radical domestic spaces. Perhaps most importantly, reworking the smart home will necessitate an understanding that domestic work is always already creative, skilled, and, under the right conditions, revolutionary.

BUILDING A SMART HOMEPLACE

One major goal of this book was to highlight the myriad ways in which technology's reach has reorganized our collective experience of home. Analyzing smart homes throughout the United States of America, I have described how smart technologies are shaping domestic relations even as they alter the physical footprint of the home. In some instances, these changes are subtle. Luxury readymade smart homes, like those encountered in chapter 1 trade on the vernacular of the suburban American dream home. Borrowing from this domestic ideal, these smart homes seek to intensify and upgrade it with automated technologies, closed-circuit surveillance, and new rooms to receive goods ordered from platformed retailers. Other smart homes, like those in chapter 2 of this text, seek to expand the definition of domesticity beyond its bourgeois ideal. These issues-oriented smart homes are living laboratories and seek to activate digital technologies to the benefit of underserved populations or political causes. Some smart homes are defined by their ability to overlay technology onto existing domestic spaces. In these retrofit smart homes, approachable platformed technology like Siri and Alexa serve as a guide to transforming the ordinary home into a smart home. No matter their configuration, these smart homes create a living record of the ways digitality is orienting, optimizing, and occupying the modern American home.

Home is a well-worn site, where, in its most just form, the grooves of care and community can be replicated on a larger scale. To some extent new domesticity is a recuperative process that highlights the value the home has always had. As a heuristic and form of political practice, new domesticity should be a recognition of the people—predominantly women, often women of color—who have made home a site of care, community, and resistance. Their work can provide a framework for reimagining the smart home as a smart homeplace. In the remainder of this chapter, I examine one such framework: *homeplace.*

In "Homeplace (a Site of Resistance)," feminist scholar and activist bell hooks demonstrates how Black women have created home as a safe space of belonging, recognition, and of collective resistance against white supremacy.[35] Creating these safe spaces was a challenge because of domination within and without. Here, homeplace is a direct response to violent, hierarchical structures that threaten people of color. Making homeplace is an imperative precisely because of these challenges. hooks describes homeplace as a site of humanization within a broader sea of dehumanizing forces. She writes that "despite the brutal reality of racial apartheid, of domination, one's homeplace was the one site where one could freely confront the issue of humanization, where one could resist. Black women resisted by making homes where all black people could strive to be subjects, not objects, where we could be affirmed in our minds and hearts despite poverty, hardship, and deprivation, where we could restore to ourselves the dignity denied us on the outside in the public world."[36] Homeplace, then, is a place of recovery, a place where the dignity of humankind is affirmed. Homeplace is a site of recognition, a place where self and other are seen and witnessed. Homeplace is a place of care and solidarity, a space of struggle toward liberation and a rejection of poverty, racism, ableism, sexism, colonization—of domination. Together, these imperatives provide a framework for imagining a new, inclusive, and socially just domesticity.

Homeplace is a useful heuristic for theorizing smart home futures and putting those theories into practice. The concept of a smart homeplace is a revolutionary concept in that it recognizes home as a site of political resistance. At the same time, it understands home as a site of care and community. These bivalent figurations are especially powerful because they demonstrate the power of the domestic sphere to interface between public and private registers. The smart homeplace is at once the site of revolutionary politics and a site of care and belonging that readies one for revolution.

The smart homeplace turns the concept of the technoliberal smart home on its head. By foregrounding a politics of inclusion, domestic technologies serve to unite and connect households rather than divide them from one another. The framework for the modern smart home draws upon hegemonic domestic traditions and folds them into the platform economy. In practice, a smart homeplace might lean into the affordances of smart technology to *connect* by connecting individual smart homes within larger communities of care and justice. A smart homeplace would leverage the networked connectivity to interlink the home with others. Rather than dividing individual households and unifying them through the platform, a smart homeplace would work to create shared domestic experiences. This might require reworking the direction of smart connectivity away from major smart tech providers and instead find ways to connect with others struggling for justice. Technically speaking, reconstructing communication patterns would likely necessitate reclaiming popular off-the-shelf smart home devices or finding ways to impede their ability to collect and monetize information. This might mean developing a parallel system of domestic technologies, away and apart from major corporate platforms. Or smart homeplaces might use the technological capacities of smart homes to meet very specific goals. Rather than using camera systems and sensors to monitor neighbors within the neighborhood (as with Ring and other smart devices), networked systems can be used to document abuses of power by police and other employees of the state.

A useful starting point for this process of building a smart homeplace would be to involve amateur home automation enthusiasts or other hacking communities. These communities offer a wealth of knowledge as well as a hacking ethos that may be instrumental to realizing such connectivity. In this configuration the home might still be an important interface for the network, but who controls it and how would change. Drawing on Kitchin and Dodge's concept of code/space, Matthew Zook and Mark Graham explore the possibility of "hacking code/space." They note that that "the functional logic of . . . code/spaces is relational" and that "individuals who must pass through these software systems for daily activities have ways of retaining agency, not only via obscuring themselves or direct sabotage but also through transgressions that subtly rework or hack code/space to their advantage."[37] One such transgression is to "find meaning via destabilising engagements with everyday systems and spaces."[38] A smart homeplace could undermine platform capitalism by rejecting its basic premises of individualist optimization, grounding the meaningful nature of the home in a care and compassion for all, rather than luxury and efficiency for a few. By virtue of its whole systems- and community-based framework, the smart homeplace can also rehabilitate the ecological mission of some smart devices, making the home sustainable in a multitude of meaningful ways. Here, the smart home can be rhetorically and spatially redefined by its ability to serve the collective rather than the self. Another form of transgression deals in the practice of hacking social and technological systems within a particular place. The project is epistemological as well as pragmatic, and community based. "This kind of hacking," they note "also prioritises information-sharing and community and focuses on understanding how systems—in this case the built environment of cities—fit together and how one gains access to parts that are normally closed off."[39] A smart homeplace may also create alternate platforms for knowledge sharing, using peer-to-peer or other formats to share this information securely.

The second major barrier to achieving the smart homeplace pertains to unequal access of home, technology, and security. Most generally, not everyone has access to the purported utopia provided by smart device products, because not everyone has access to safe housing in general. The future of the smart home must therefore be accountable to unhoused populations, contingently housed populations, and others without housing stability. One productive way forward would be to align the smart homeplace with the housing justice movement. The housing justice movement argues that social, political, and economic inequalities are borne out in housing practices. These practices, propelled by capitalism, racism, settler colonialism, and imperialism, create stark divisions between "the housing landscapes of the wealthy and the housing crisis experienced by marginalized communities."[40] Ananya Roy locates housing inequality within broader structures of domination, encouraging us to "situate the housing question in racial capitalism and its logics of dispossession and subordination.[41]" Housing justice demands structural transformation, including a rethinking of fundamental capitalist logics, such as ownership of land, personhood, and property.[42]

Such an approach calls into question many fundamental aspects of smart home discourses. Applying a housing justice lens to the smart home would mean rejecting the smart home's normative, bourgeois conceptions of home in favor of a collective, decommodified, and radical praxis. Alternate frameworks for housing may be more suitable to meet the varying needs of the population. Sophia Maalsen, a smart housing scholar, has examined how smart technologies can support shared housing.[43] "Projects such as co-living and platform-enabled housing services," while imperfect, "offer potential for crucial intervention in the corporate smart city and investment driven housing markets."[44] Co-living and shared housing can be a useful alternative to standalone housing, although the distribution of different types of housing is likely to reify difference without adequate forethought and oversight. A housing justice approach echoes the framework of smart homeplace, grounding agency and empowerment in people and communities, rather than in housing markets or technocapitalists.

A smart homeplace recognizes that domesticity is made possible by workers performing what is often underpaid and undervalued domestic labor—inside and outside of the home. Building a smart homeplace must therefore operate in solidarity with labor movements, particularly those which press major technology companies to be transparent with their labor practices and subsequently hold them accountable for labor violations. The United States is experiencing a new wave of labor activism that can prove instructive for imagining social relations at home and in public. For instance, in spring of 2022, as the full impacts of COVID-19 on frontline workers came into focus, workers at the Staten Island Amazon warehouse voted to unionize. Their actions have been heralded as a template for activating workers across the technology and logistics industry. Writing for *The Atlantic*, Adam Serwer argues that "the story of the campaign is almost cinematic: workers hailing from all around the globe running a multilingual campaign to defeat a corporate behemoth that subjects its wage workers to harrowing conditions."[45] According to Serwer, one of the most important elements of the Staten Island workers' strategy was calling out Amazon's attempts to position itself as "egalitarian" or "enlightened." Pointing out the radical inconsistencies between Amazon's corporate image campaign and working conditions on the ground helped cut through the noise that might undermine collective action. Moreover, these workers, fronted in part by Christian Smalls, had to overcome Amazon's repeated attempts to frame the Black man as "not smart or articulate" and therefore incapable of leading a movement for labor rights.[46]

Amazon Labor Union workers' agitation toward labor justice—and Amazon's intense and possibly illegal resistance to these efforts—treads on familiar ground in our investigation of smart homes. It should not go unnoticed that Amazon's strategy of undermining unionization mobilizes white supremacist and eugenicist conceptions of intelligence to delegitimize a racialized subject for a (presumably white) audience, as these very same conceptions of smartness architect the justification for the smart home. Amazon's corporate positioning around social justice issues attempted to incorporate dissent rather than respond to it. Living in digitality's expansive and incorporative framework supplies the means for platform hegemons to sidestep critique by promising a utopian future only it controls. In this future a handful of privileged digital subjects experience ease or comfort in exchange for the continued subordination of working classes. But it will not be long until the only privileged subjects left are the platform capitalists themselves.

A more just smart home future must recognize and resist global flows of capital that incentivize outsourcing operations to countries with less stringent labor and environmental practices. The benefits of smart homes are acquired on the backs of invisible others, many of whom experience domination and marginalization. We saw, for instance, how platform laborers are rhetorically and spatially separated from the benefits of the smart environments for which they labor, even as they are extensively surveilled, tracked, and monitored to uphold impossible efficiency standards. Manufacturing smart devices also takes a toll—on the earth in the form of rare earth minerals and on the people who create the devices. As Hilary Bergen reports, "working conditions at Foxconn, where many of Apple's products are made and assembled, are so dire that in 2010, eighteen workers attempted suicide, resulting in fourteen deaths."[47] These individuals labor in inhumane environments with near machinic skill and efficiency. They are paid a pittance to produce enormous value. In the fictive utopia depicted by smart home discourses, these laborers are almost never visible even though they are central to the experience of living in digitality. Indeed, "the abused and exploited bodies at Foxconn make up one branch of the new digital cyborg assemblage, but are rendered invisible in the final product of our iPhone. . . . The Foxconn workers, machine-like and imprisoned, forced to perform not only technical but affective labor, are much like the virtual cyborg herself."[48] The polished, gendered figures of Siri and Alexa are presented instead of the actual laborers who make smart homes possible.

The smart home experience, as it is currently imagined, is only the modern expression of a vast sociotechnical process that could continue unencumbered without mobilized response. Much of the dominant smart home framework hides the supply-side logistics, services, and labor needed to provide a seamless smart home experience. This process is constituted by human and nonhuman actors that form the assemblage that is the smart home. We must make visible their contributions to the process that is making the smart home. That means highlighting the workers who mine rare earth materials and work in device manufacturing plants, and showcasing the designers, engineers, and other white- and no-collar laborers who imagine the future of the smart home. It also means standing in solidarity with workers on strike and those attempting to unionize platform workplaces—in Silicon Valley and beyond.[49]

The third step I propose here to build a more just smart homeplace is decidedly more abstract. We must imagine and incorporate new domestic ideals that break down frameworks of domination rather than propping them up. Guided by bell hooks's theorization of homeplace as a site of belonging, I offer below several "gathering points" that can direct our future efforts. It is my hope that these gathering points contribute shared values to larger, ongoing conversations about the future of living in digitality within and beyond the smart home.

- The smart homeplace is a site of belonging where we hang together by virtue of our sameness and difference, where our value is not in what we can produce, or what we buy, or how we fit into a system not built for us but instead in a recognition of and celebration of our shared humanity. The smart homeplace rejects exclusionary, colonialist, and ableist definitions of intelligence, favoring instead a tactical recognition of knowledges and knowledge practices that have been silenced, ostracized, or disciplined as other. The accumulation of capital in the here and now is not the standard of living with technology. Timelines grow longer, accumulation is traded for abundance, rugged individualism is traded for careful and persistent collectivity, and the domestic sphere shifts to incorporate these new intelligences.
- The smart homeplace is a site of refuge, where we come together to reflect on our respective struggles and prepare for the next day together. Here, the home is not a fortress where we are alienated from others, or where we divide ourselves from people who are not like us, but a place where we pour into each other the good things that make us unique. It is a place full of food, drink, culture, love, feeling, sensation, pleasure, and comfort. It is also a place where we can hold ourselves and others accountable to larger goals, such as challenging dehumanization and domination endemic to living in digitality,

including platform capitalism and technoliberalism. This place—between ease/comfort and the ongoing, active labor of care—provides the energy necessary for the smart homeplace—and the people in it—to thrive outside capitalist metrics.

- The smart home homeplace is at once spatial and "placed" and transcendent and "open"—it is a place that forms when we are among intimates, and when we care for others. We can take it with us, and we can return to it. It is networked within and beyond the technological sphere. It is imaginary and materially accessible. It recognizes that family is an expansive condition rather than a set of predetermined social relations. Having done so the smart homeplace challenges domestic technologies that rely on or reinforce exclusionary principles of home, self, other, kin.
- The smart homeplace and the people who occupy it are consistently self-reflexive of the value and import of domesticity, as a site of labor, as a site of relation spatialized in the (smart) home, and as a place of creation and care. The smart homeplace recognizes that the domestic sphere has typically been assigned the designation of "feminine." The smart homeplace challenges normative heterosexism, while also recognizing the values typically assigned to the feminine—care, collaboration, community, creation. It does not seek to recuperate the values dominant in the worldmaking of present digitality; rather, it seeks to remake the technology by attending to and transforming the values toward care, community, and shared placemaking.
- The smart homeplace recognizes that language and imagination become actions, processes, and places. As such, the smart homeplace revels in and supports what bell hooks calls the "will to meaning." For hooks, "when the will to meaning is paramount, human life retains dignity. The capacity of humans to create community, to make connections, to love, is nurtured and sustained."[50] People who occupy the smart homeplace understand that language matters, and that the smartest homes deploy technology such that humans retain dignity, make connections, nurture, and love. Technology—as a set of objects, infrastructures, and meanings—is refigured within the realm of what Padilla Carrol calls "new domesticity," which refracts the value of home as a space of worldmaking. Divisions between self and other, articulated and affirmed through the spatial relations of hegemonic smart domesticity, are recalibrated through language determined to incorporate not just privileged bodies but all bodies. Privileged digital subjects abdicate luxuries afforded to them by the smart homes in exchange for a new abundance. Technology's

efficiencies and profitabilities, which are presently enjoyed by a concentrated few neo–Gilded Age elites, can instead create opportunities for ease and leisure for the masses, rather than the few.

- The smart homeplace uses emerging technology in service of care, inclusion, solidarity, and dignity. When necessary, the smart homeplace reworks or hacks smart technologies to hold platform capitalists and their partners accountable. Whereas surveillance apparatuses typically place power in the hands of the individual, the corporation, or institutions purporting to police these, the smart homeplace technology recognizes that emerging technologies can be hacked to "watch the watchers." The smart homeplace becomes a site for political activation, using technologies where prudent and eschewing them when necessary.
- When possible the smart homeplace builds alternative technological networks, linking social networks together for knowledge sharing, creation, and support. Often the smart homeplace imagines alternate frameworks of knowledge and care, linking self and others through the literal and figurative creation of networked infrastructures and technologies. The smart homeplace recognizes the communities and collectives that have been doing this work already in visible and invisible ways. It finds ways to cultivate connections between activist and advocacy organizations with the skills and capacity to mount a challenge to the technoliberal institution at all levels.
- The smart homeplace architects a new set of relationships and ways of being with others from the domestic sphere. In recognition of the worldmaking and world-breaking capacities of domesticity, the smart homeplace seeks to spatialize core values of collectivity, care, and abundance inside the home and outside of it.

The home has never been—nor can it be—a space "set off" from the realm of social relations. Nor is it separate from the technical infrastructure that necessitated its formation. Home is not an island, discrete and separate from the world. It is not "an autonomous entity." Rather, it is a site of social relation and therefore operates in the realm of publicity even as it is private. Any attempt to rehabilitate or rework the smart home must consider its multiplicity, its paradoxical nature, and its meaningful role in society.

The shift toward digitality—toward ubiquitous and ambient computing—paired with the expansive capacity of the platform, has created conditions under which the home was recast in the image of platform capitalism. But this cultural schism is also productive for projects imagining otherwise. In the past, homeplace has been the product of the domestic labor of people on the margins. For smart homeplace to be functional at any level—pragmatic or imaginary—people with privilege will need to actively participate in the process of its (un)making. This will likely mean divesting in the systems of privilege and power that smart homes currently operationalize to sell smart home products. It will be impossible to build a smart homeplace for all if people in positions of power continue to buy into the framework that the smart home is apolitical or unbiased. It will also be impossible to build a smart homeplace that is inclusive and egalitarian if individuals continue to use smart devices as they are intended.

The case studies in this book evidence that smart homes often operate as a shadow of homeplace. The mass-mediated experience of living in digitality offered by modern smart homes traffics in language of homeplace but does not fulfill its most basic promises. It perverts them. Smart home discourses purport to provide the experience of homeplace. Smart homes are sold with the promise of ease and choice and agency. They purport to increase leisure time, to alleviate repetitive labor, even to care for their residents. Like a true homeplace, the smart home is figured as the antithesis of domination and deprivation and hardship. Buying products will ease our anxiety, make our lives run smoother, help us be more present in our relationships. In some instances, smart homes can achieve these goals for their homeowners. But they come at a great cost to both smart home users and others within the broader smart home assemblage.

The potential of the smart home is poisoned by technocratic frameworks that justify and formalize domination using the language of traditional domesticity. The fundamental components of home are reworked until they are no longer recognizable as such. Caring is subsumed by platformed data extraction and monitoring. Community is recognizable only within the infrastructure of the platform. Ease and efficiency are possible only if one is willing to turn the other cheek to the violence that makes the smart home possible. Human dignity for some is traded for accelerating profit for others. The revolutionary power of place is domesticated—literally and figuratively—by platform capitalists.

Usefully, as a theoretical framework and as an experience of space/collectivity, the smart homeplace is well positioned to respond to dominant structures and violence endemic to living in digitality. Indeed, the subversive nature of homeplace is gained in part through its opposition to violent, hegemonic reality structures. The smart homeplace provides the physical space to oppose the platform economy even as it foregrounds the care necessary to continue the fight. As a site of recovery, it rejects outright the premise that humankind's value is in its ability to be monitored, tracked, and monetized in the digital sphere. Rather, the smart homeplace creates literal and figurative space to struggle against technocratic politics that organize the home and the spaces beyond it.

We can remake the figure of the smart home by remaking the homeplace. A feminist approach to home is driven by principles antithetical to the technoliberal smart home. Feminist approaches to domesticity by necessity call into question the very precepts of capitalism as an economic system and living in digitality as an organizing principle by asking how value manifests in the home.[51] By foregrounding a smart homeplace, we can recalibrate that value toward human dignity, not profit. The task is urgent. In beginning to build the future, there is no time like the present.

Notes

Preface

1. Derek Thompson, "Go Home Now," *The Atlantic*, December 1, 2020.

2. Sophia Maalsen and Robyn Dowling, "Covid-19 and the Accelerating Smart Home," *Big Data & Society* 7, no. 2 (July 2020).

3. Maalsen & Dowling, "Covid-19 and the Accelerating," 2.

4. Maalsen & Dowling, "Covid-19 and the Accelerating," 4.

5. Maalsen & Dowling, "Covid-19 and the Accelerating," 1, 4.

6. Amy G. Richter, *At Home in Nineteenth-Century America: A Documentary History* (New York: NYU Press, 2015), 3; Ruth Schwartz Cowan, *More Work for Mother: The Ironies of Household Technology from the Open Hearth to the Microwave*, 2nd ed. (New York: Basic Books, 1985), 38; H. Ward Jandl, "Introduction," in *Yesterday's Houses of Tomorrow: Innovative American Homes, 1850 to 1950* (Washington, DC: Preservation Press, 1991), 5.

7. Amy G. Richter, *At Home in Nineteenth-Century America*, 1–2.

8. Importantly, one needn't be physically outside of the home to be rendered as threat to the domestic space. As Katherine McKittrick notes, "geographies of whiteness . . . are, for the most part, rendered protectable. Slave quarters, plantation homes, fields, kitchens, are, for black women unprotected . . . the body is rightfully retranslated as inferior, captive, and accessible to violence." Katherine McKittrick, *Demonic Grounds: Black Women and the Cartographies of Struggle* (Minneapolis: University of Minnesota Press, 2006), 82.

9. As Richter notes, the domestic ideal was both conservative and radical in its connection to morality and citizenship: "Recognizing the ways in which domestic order and morality had become credentials for citizenship and objects of public policy, many marginalized groups seized upon them to advance a variety of political agendas." Richter, *At Home in Nineteenth-Century America*, 97.

10. Cowan, *More Work for Mother*, 38.

11. Bruce Mitchell and Juan Franco, "HOLC 'Redlining' Maps: The Persistent Structure of Segregation and Economic Inequality" (NCRC, March 2018); Katie Lambright, "Digital Redlining: The Nextdoor App and the Neighborhood of Make-Believe," *Cultural Critique* 103 (2019): 84–90; Safiya Umoja Noble, *Algorithms of Oppression: How Search Engines Reinforce Racism* (New York: NYU Press, 2018), 1.

12. Davin Heckman, *A Small World: Smart Houses and the Dream of the Perfect Day* (Durham, NC: Duke University Press, 2008).

13. Sarah J. Darby, "Smart Technology in the Home: Time for More Clarity," *Building Research & Information* 46, no. 1 (January 2, 2018): 140–41.

14. Lynn Spigel, *Make Room for TV: Television and the Family Ideal in Postwar America* (Chicago: University of Chicago Press), 1992, 11.

15. Zaheer Baber, "Whitey on the Moon: NASA and Pressing Social Movements," *Times Literary Supplement*, September 1, 2017.

16. Some, like the Southern Christian Leadership Conference president the Reverend Ralph Abernathy lamented that these technological innovations represented a "distorted sense of the national priorities" indicative of a "tragic and inexcusable gulf that exists between America's technological abilities and social injustices." As quoted in Zaheer Baber, "Whitey on the Moon: NASA and Pressing Social Movements," *Times Literary Supplement*, September 1, 2017.

17. Maureen Ryan, "*Entertaining* Fantasies: Lifestyle and Social Life in 1980s America," *Journal of Communication Inquiry* 39, no. 1 (2015): 85.

18. As Maureen Ryan notes, "neoconservatism was . . . a way of organizing and legitimizing these neoliberal economic principles, rejecting the welfare state and . . . anchoring the chaotic potential of market freedom with a moral foundation in Christian, family-oriented values." Ryan, "*Entertaining* Fantasies," 85.

19. As reported in Mike Berry et al., "How Smart Is 'Smart'? Smart Homes and Sustainability," in *Steering Sustainability in an Urbanizing World: Policy, Practice and Performance*, ed. Anitra Nelson (Hampshire, Burlington: Ashgate, 2007), 240.

20. Berry et al., "How Smart Is 'Smart'?," 240.

21. For instance, by the middle of 2020, the "Zoom room" became trendy in the world of real estate. The term refers to small private or semiprivate spaces where individuals can join virtual meetings via Zoom, a web conferencing platform-turned-verb. Dawson White, "Homes Listings with Zoom Rooms on the Rise, Zillow Says," *Kansas City Star*, January 5, 2021; Andrew Chamings, "The 'Zoom Room' Boom: A Very 2020 Real Estate Trend," *SFGATE*, July 24, 2020.

22. Taylor Locke, "From Banana Bread to Lasagna: Top 10 Most-Googled Recipes while People Are Self-Quarantining," CNBC, April 26, 2020.

23. "'Victory Gardens' for the War against COVID-19," *CBS News*, April 5, 2020.

24. Fred Imbert, "Dow Drops Nearly 3,000 Points, as Coronavirus Collapse Continues; Worst Day since '87," CNBC, March 15, 2020.

25. Imbert, "Dow Drops."

26. "US Oil Futures Plunge below $0 for First Time in History," *Al Jazeera*, April 27, 2020.

27. Sean Szymkowski, "Lowest Gas Prices in the US Falls [*sic*] to Just 91 Cents a Gallon," *CNET Roadshow*, accessed January 25, 2021.

28. Lance Lambert, "Real Unemployment Rate Soars Past 20%—and the U.S. Has Now Lost 26.5 Million Jobs," *Fortune*, April 23, 2020.

29. US Census Bureau, "Income and Poverty in the United States: 2020," Census.gov, accessed June 15, 2022.

30. Fabian Mendez Ramos and Jaime Lara, "COVID-19 and Poverty Vulnerability," *Brookings* (blog), May 18, 2022.

31. Samuel Stebbins and Grant Suneson, "Jeff Bezos, Elon Musk among US Billionaires Getting Richer during Coronavirus Pandemic," *USA TODAY*, December 1, 2020, accessed January 25, 2021.

32. Samuel Stebbins and Grant Suneson, "Jeff Bezos, Elon Musk."

33. CDC and HHS, "Temporary Halt in Residential Evictions to Prevent the Further Spread of COVID-19," Federal Register, September 4, 2020.

34. According to the CDC and HHS, "housing stability helps protect public health

because homelessness increases the likelihood of individuals moving into congregate settings, such as homeless shelters, which then puts individuals at higher risk to COVID-19. . . . Unsheltered homelessness also increases the risk that individuals will experience severe illness from COVID-19." See CDC and HHS, "Temporary Halt in Residential Evictions to Prevent the Further Spread of COVID-19," Federal Register, September 4, 2020.

See also CDC, "Interim Guidance on Unsheltered Homelessness and Coronavirus Disease 2019 (COVID-19) for Homeless Service Providers and Local Officials," Centers for Disease Control and Prevention, August 6, 2020.

35. Keya Vakil, "'Those Numbers Take Your Breath Away': Why Black Americans Are Dying from COVID-19 at Alarming Rates," Courier, April 20, 2020.

36. Neeta Thakur et al., "The Structural and Social Determinants of the Racial/Ethnic Disparities in the U.S. COVID-19 Pandemic: What's Our Role?," *American Journal of Respiratory and Critical Care Medicine* 202, no. 7 (October 2020): 943.

37. World Health Organization, "WHO Coronavirus (COVID-19) Dashboard."

Introduction

1. Patrick Lucas Austin, "What Will Smart Homes Look like in 10 Years?" *Time*, July 25, 2019.

2. Dieter Bohn, "Exclusive: Amazon Says 100 Million Alexa Devices Have Been Sold," The Verge, January 4, 2019.

3. Rishi Chandra and Scott Huffman, "How Google Home and the Google Assistant Helped You Get More Done in 2017," *Google* (blog), January 5, 2018.

4. Jaeyeon Jung, as quoted in "Samsung Electronics Expands Its Galaxy Upcycling Programme, Enabling Customers to Repurpose Galaxy Smartphones into Smart Home Devices," *Samsung News* (blog), accessed September 24, 2021.

5. Sophia Maalsen and Robyn Dowling, "Covid-19 and the Accelerating Smart Home," *Big Data & Society* 7, no. 2 (July 2020): 205395172093807.

6. Dawson White, "'Zoom Rooms?' They're among the Top Home Trends for 2021, Zillow Says," *Kansas City Star*, January 5, 2021.

7. Maalsen and Dowling, "Covid-19 and the Accelerating Smart Home," 2.

8. The good life is a rhetorical construction that manifests spatially. For an extended treatment of the good life, particularly in the context of suburbia, see Greg Dickinson, *Suburban Dreams: Imagining and Building the Good Life* (Tuscaloosa: University of Alabama Press, 2015).

9. Yolande Strengers, *Smart Energy Technologies in Everyday Life: Smart Utopia?*, (Dordrecht: Springer, 2013), 2.

10. Germaine R. Halegoua, *Smart Cities* (Cambridge: MIT Press, 2020).

11. Sophie Davies. "Wi-Fi but No Water: Can Smart Tech Help a City's Poor?" *Reuters*, January 5, 2018.

12. "Amazon Apologises for Wrongly Denying Drivers Need to Urinate in Bottles," *BBC News*, April 4, 2021.

13. Yolande Strengers and Jenny Kennedy. *The Smart Wife: Why Siri, Alexa, and Other Smart Home Devices Need a Feminist Reboot* (Cambridge: MIT Press, 2020).

14. Sarah Sharma makes a similar claim in her treatise on the formation of digital subjects operating within unequal temporal infrastructures. Sarah Sharma, *In the Meantime: Temporality and Cultural Politics* (Durham, NC: Duke University Press, 2014).

15. Paul N. Edwards, "Infrastructure and Modernity Infrastructure and Modernity: Force, Time, and Social Organization in the History of Sociotechnical Systems," in *Modernity and Technology*, edited by Thomas J. Misa, Philip Brey, and Andrew Feenberg (Cambridge: The MIT Press, 2003), 186.

16. My goal here is not to be exhaustive in my characterization of digitality as an ideological and spatial formation, nor is it to definitively characterize the contours of the smart home. In fact, this heuristic is necessarily incomplete, because its object is perpetually unfolding and because my thinking is influenced by my position as a privileged (digital) subject. My interest instead is building on the insights of other thinkers and advocates in laying the foundation for an ongoing scholarly and public-facing conversation about how life is and ought to be lived with ubiquitous, immersive technology.

17. Seb Franklin, *Control: Digitality as Cultural Logic* (MIT Press, 2015), xix.

18. Shannon Mattern. "A City Is Not a Computer." *Places Journal*, February 7, 2017.

19. I borrow the term "protocological" from Damien Smith Pfister, who writes that "Protocological fascism identifies the propensity for digital media dominated by capitalism to collapse *logos* into logistics, standardizing interactions through protocols . . . the concept of protocological fascism steers critical attention to how the enabling infrastructures of digitality are created through industrial processes and undemocratic procedures." Damien Smith Pfister, "Digitality, Rhetoric, and Protocological Fascism; Or, Fascist Ants & Democratic Cicadas," *Journal for the History of Rhetoric* 23, no. 1 (2020): 4.

20. Galen Gruman, as cited in Yolande Strengers and Jenny Kennedy, *The Smart Wife*, 28.

21. Larissa Nichols and Yolande Strengers, "Robotic Vacuum Cleaners Save Energy? Raising Cleanliness Conventions and Energy Demand in Australian Households with Smart Home Technologies," *Energy Research & Social Science* 50 (April 1, 2019): 73–81.

22. As Harney and Moten have written, "And anyway for capital the subject has become too cumbersome, too slow, too prone to error, too controlling, to say nothing of too rarified, too specialized a form of life. Yet it is not we who ask this question. This is the automatic, insistent, driving question of the field of logistics. Logistics wants to dispense with the subject altogether. This is the dream of this newly dominant capitalist science." Stefano Harney and Fred Moten, *The Undercommons: Fugitive Planning & Black Study* (Wivenhoe, NY: Minor Compositions, 2013), 87.

23. Yolande Strengers, *Smart Energy Technologies in Everyday Life: Smart Utopia?* (Dordrecht: Springer, 2013), 1–2.

24. Strengers, *Smart Energy Technologies*, 2.

25. Niccolò Cuppini and Mattia Frapporti, "Logistics Genealogies: A Dialogue with Stefano Harney," *Social Text* 36, no. 3 (September 1, 2018): 95–110; Harney and Moten, *Undercommons*.

26. Stefano and Moten, *Undercommons*, 92.

27. Stefano and Moten, *Undercommons*, 91.

28. Nick Srnicek, *Platform Capitalism* (Cambridge, UK: Polity, 2016); Nick Couldry and Ulises Ali Mejias, *The Costs of Connection: How Data Is Colonizing Human Life and Appropriating It for Capitalism* (Stanford, CA: Stanford University Press, 2019).

29. Couldry and Mejias, *Costs of Connection*, 5.

30. Couldry and Mejias, *Costs of Connection*, 5.

31. Sharma, *In the Meantime*, 19.

32. Srnicek, *Platform Capitalism*, 128.

33. Srnicek, *Platform Capitalism*, 44.

34. Bohn, "Exclusive: Amazon Says."

35. Rob Kitchin and Martin Dodge, *Code/Space: Software and Everyday Life* (Cambridge: MIT Press, 2011), 159–60.

36. Murray Goulden, "'Delete the Family': Platform Families and the Colonisation of the Smart Home," *Information, Communication & Society* (2019): 3.

37. Maalsen draws on the philosophy of Deleuze and Guattari to make a related claim about the smart home as an assemblage. Maalsen, "Revising the Smart Home as Assemblage," *Housing Studies* (2019): 1–16.

38. Franklin, *Control*, 85.

39. In her examination of smart homes, Yolande Strengers describes the formation of a novel "energy consumer who is intended to both realise and significantly benefit from this vision" of a smart utopia. She calls her new consumer "Resource Man." *Smart Energy Technologies*, 2.

40. Spigel, *Make Room for TV*; Lynn Spigel, *Welcome to the Dreamhouse: Popular Media and Postwar Suburbs* (Durham, NC: Duke University Press, 2001); Ruth Schwartz Cowan, *More Work for Mother: The Ironies of Household Technology from the Open Hearth to the Microwave* (New York: Basic Books, 1985).

41. Alain de Botton, *The Architecture of Happiness* (New York: Vintage Internat, 2008), 137.

42. de Botton, *Architecture of Happiness*, 78.

43. Franklin, *Control*, xv.

44. Franklin, *Control*, xix.

45. Franklin, *Control*, xix.

46. Franklin, *Control*, xv.

47. Franklin, *Control*, 85.

48. Couldry and Mejias, *Costs of Connection*, 5.

49. Couldry and Mejias make a strong argument for data colonialism, or "an emerging order for the appropriation of human life so that data can be continuously extracted from it for profit." *Costs of Connection*, xiii.

50. Couldry and Mejias, *Costs of Connection*, xviii.

51. Franklin, *Control*, 83.

52. They write that the "self's minimal integrity is the boundedness that constitutes a self *as a self*. Often that boundedness is experienced in defending a minimal space of physical control around the body, but it can also be invaded without physical incursion—for example, by acts of power that intimidate, shame, harass, and monitor the self." Couldry and Mejias, *Costs of Connection*, 158, italics original.

53. Couldry and Mejias, *Costs of Connection*, 5.

54. Couldry and Mejias, *Costs of Connection*, 5, italics added.

55. Couldry and Mejias, *Costs of Connection*, 159.

56. Couldry and Mejias, *Costs of Connection*, 159–60.

57. Franklin, *Control*, xv.

58. Franklin, *Control*, xvi.

59. Franklin, *Control*, 7.

60. Jathan Sadowski, *Too Smart: How Digital Capitalism Is Extracting Data, Controlling Our Lives, and Taking over the World* (Cambridge: MIT Press, 2020), 9.

61. Joan Faber McAlister, "Figural Materialism: Renovating Marriage through the American Family Home," *Southern Communication Journal* 76, no. 4 (September 1, 2011): 279–304; Lauren Berlant and Michael Warner, "Sex in Public," *Critical Inquiry* 24, no. 2 (1998): 547–66; Hilde Heynen and Gulsum Baydar, eds., *Negotiating Domesticity: Spatial Productions of Gender in Modern Architecture* (London: Routledge, 2005); Christine Atha, "Dirt and Disorder: Taste and Anxiety in the Homes of the British Working Class," in *Atomic Dwelling: Anxiety, Domesticity, and Postwar Architecture*, ed. Robin Schuldenfrei, 207–26 (Abingdon, Oxon: Routledge, 2012).

62. Safiya Umoja Noble, *Algorithms of Oppression: How Search Engines Reinforce Racism* (New York: NYU Press, 2018), 31.

63. André L. Brock, *Distributed Blackness: African American Cybercultures* (New York: New York University Press, 2019), 16.

64. Andrew Feenberg, "What Is Philosophy of Technology" (Lecture for Komaba Undergraduates, Simon Fraser University, 2003).

65. Brock, *Distributed Blackness*, 17.

66. Sut Jhally, featuring an interview with bell hooks. *Media Education Foundation Transcript, bell hooks: Cultural Criticism & Transformation*, 1997.

67. Sadowski, *Too Smart*, 10.

68. Amy Kaplan, "Manifest Domesticity," *American Literature* 70, no. 3 (1998): 594, 560.

69. Amy Kaplan has argued that domesticity was historically projected in a nationalistic and moralistic context: "Part of the cultural work of domesticity might be to unite men and women in a national domain and to generate notions of the foreign against which the nation can be imagined as home. . . . Through the process of domestication, the home contains within itself those wild or foreign elements that must be tamed; domesticity not only monitors the borders between the civilized and the savage but also regulates traces of the savage within itself." Kaplan, "Manifest Domesticity," 582.

70. Hortense J. Spillers, *Black, White, and in Color: Essays on American Literature and Culture* (Chicago: University of Chicago Press, 2003), 178.

71. Couldry and Mejias, *Costs of Connection*, xvii.

72. Katherine McKittrick, "Mathematics Black Life," *Black Scholar* 44, no. 2 (Summer 2014): 21.

73. Von Diaz, "La Piña: A Pineapple Rum Inspired an Exquisite Daiquiri—and Reflections on History," *Garden & Gun*, August/September 2021.

74. Thao Phan, "Amazon Echo and the Aesthetics of Whiteness," *Catalyst: Feminism, Theory, Technoscience* 5, no. 1 (2019).

75. Bruce Mitchell and Juan Franco, "HOLC 'Redlining' Maps: The Persistent Structure of Segregation and Economic Inequality" (NCRC, March 2018).

76. Damien Smith Pfister, "Technoliberal Rhetoric, Civic Attention, and Common Sensation in Sergey Brin's 'Why Google Glass?'" *Quarterly Journal of Speech* 105, no. 2 (2019): 182–203.

77. Terri Friedline and Zibei Chen, "Digital Redlining and the Fintech Marketplace: Evidence from US Zip Codes," *Journal of Consumer Affairs* 55, no. 2 (2021): 370.

78. Friedline and Chen, "Digital Redlining," 369. In her book *Algorithms of Oppression*, Safiya Umoja Noble advances the similar concept of *technological redlining* to describe how algorithms' "digital decisions reinforce oppressive social relationships and enact new modes of racial profiling," 1.

79. Virginia Eubanks, *Automating Inequality: How High-Tech Tools Profile, Police, and Punish the Poor* (New York: Picador St. Martin's Press, 2019), 8.

80. Torin Monahan, *Surveillance in the Time of Insecurity* (New Brunswick, NJ: Rutgers University Press, 2010), 82.

81. Monahan, *Surveillance in the Time of Insecurity*, 81.

82. Katie Lambright, "Digital Redlining: The Nextdoor App and the Neighborhood of Make-Believe," *Cultural Critique* 103 (2019): 86.

83. See chapter 1 for a robust discussion of infrastructure relative to domesticity.

84. Niels van Doorn, "Platform Labor: On the Gendered and Racialized Exploitation of Low-Income Service Work in the 'On-Demand Economy,'" *Information, Communication & Society* 20, no. 6 (2017): 898–914.

85. Phan, "Amazon Echo and the Aesthetics of Whiteness"; Heather Suzanne Woods, "Asking More of Siri and Alexa: Feminine Persona in Service of Surveillance Capitalism," *Critical Studies in Media Communication* 35, no. 4 (2018): 334–49.

86. See Strengers on *Resource Man*, the energy-consumptive subject rendered by technoutopian smart homes. *Smart Energy Technologies*, 2.

87. Jathan Sadowski, Yolande Strengers, and Jenny Kennedy, "More Work for Big Mother: Revaluing Care and Control in Smart Homes," *Environment and Planning A: Economy and Space*, June 14, 2021, 0308518X211022366, 3.

88. Phan, "Amazon Echo and the Aesthetics of Whiteness."

89. Maalsen, "Revising the Smart Home as Assemblage," 1.

90. Darby, "Smart Technology in the Home," 141.

91. Sadowski, Strengers, and Kennedy, "More Work for Big Mother," 2.

92. Kitchin and Dodge, *Code/Space*, 159.

93. Sadowski, *Too Smart*, 27.

94. Mark Andrejevic, "Foreword," in *Feminist Surveillance Studies*, ed. Rachel E. Dubrofsky and Shoshana Magnet (Durham, NC: Duke University Press, 2015).

95. Srnicek, *Platform Capitalism*.

96. CNBC Television, *Watch CNBC's Exclusive Interview with Amazon's SVP of Devices David Limp*, 2019.

97. Strengers, *Smart Energy Technologies*, 23.

98. Srnicek, *Platform Capitalism*; Sebastian Olma, "Never Mind the Sharing Economy: Here's Platform Capitalism," Institute of Network Cultures, October 16, 2014.

99. Maalsen, "Revising the Smart Home."

100. Strengers, *Smart Energy Technologies*, 26.

101. Strengers, *Smart Energy Technologies*, 27.

102. Jessica Enoch, *Domestic Occupations: Spatial Rhetorics and Women's Work*. Studies in Rhetorics and Feminism (Carbondale: Southern Illinois University Press, 2019), 9.

103. Sandy Isenstadt, *The Modern American House: Spaciousness and Middle Class Identity* (Cambridge: Cambridge University Press, 2006), 11.

104. Enoch, *Domestic Occupations*.

105. Enoch, *Domestic Occupations*, 6.

106. Enoch, *Domestic Occupations*, 6.

107. Dickinson, *Suburban Dreams*, 5.

108. Dickinson, *Suburban Dreams*, 5.

109. Germaine R. Halegoua, *The Digital City: Media and the Social Production of Place* (New York: NYU Press, 2020), 14.

110. Halegoua, *Digital City*, 15.

111. Candice Rai and Caroline Gottschalk Druschke, "On Being There: An Introduction to Studying Rhetoric in the Field," in *Field Rhetoric: Ethnography, Ecology, and Engagement in the Places of Persuasion*, ed. Candice Rai and Caroline Gottschalk Druschke, 1–21. (Tuscaloosa: University of Alabama Press, 2018). Scholars in communication recognize the transformative rhetoricity of tours. Pezzullo, for instance, argues that tours are not only rhetorical but potentially transformative insofar as they "rhetorically invent more livable designs for life." For Pezzullo, tours are performance events; "although tour guides . . . may have written themselves scripts," tours are not only descriptive of a community, place, or partnership. Rather, tours can be understood as thoughtful public presentations of self and community. Tours, then, "are constitutive rather than 'fake,'" despite scripted performances. Phaedra C. Pezzullo, "Touring 'Cancer Alley,' Louisiana: Performances of Community and Memory for Environmental Justice," *Text and Performance Quarterly* 23, no. 3 (2003): 228.

112. Yolande Strengers Jenny Kennedy, Paula Arcari, Larissa Nicholls, and Melissa Gregg, "Protection, Productivity and Pleasure in the Smart Home: Emerging Expectations and Gendered Insights from Australian Early Adopters," *ACM*, 2019, 645.

113. Davin Heckman, *A Small World: Smart Houses and the Dream of the Perfect Day* (Durham, NC: Duke University Press, 2008), 14–15.

114. Kitchin and Dodge, *Code/Space*, 17.

115. Kirsten Gram-Hanssen and Sarah J. Darby, "'Home Is Where the Smart Is'? Evaluating Smart Home Research and Approaches against the Concept of Home," *Energy Research & Social Science* 37 (March 2018): 97.

116. bell hooks, *Yearning: Race, Gender, and Cultural Politics* (Boston: South End Press, 1999), 47.

Chapter 1

1. Lynn Spigel, *Make Room for TV: Television and the Family Ideal in Postwar America* (Chicago: University of Chicago Press, 1992), 3. Although other scholars, including Lauren Heintz and Kathleen McHugh have used the term "domestic futurity," I am, to my knowledge, the first at advancing this particular definition.

2. Niels van Doorn, "Platform Labor: On the Gendered and Racialized Exploitation of Low-Income Service Work in the 'On-Demand' Economy," *Information, Communication & Society* 20, no. 6 (2017): 898–914.

3. Gulsum Baydar, "Figures of Wo/Man in Contemporary Architectural Discourse," in *Negotiating Domesticity: Spatial Productions of Gender in Modern Architecture*, ed. Hilde Heynen and Gulsum Baydar (London: Routledge, 2005), 30.

4. Paula Baker, "The Domestication of Politics: Women and American Political Society, 1780–1920," *American Historical Review* 89, no. 3 (1984): 620.

5. I borrow Spigel's terminology of "competing paradigms" in media homes. Lynn Spigel, *Welcome to the Dreamhouse: Popular Media and Postwar Suburbs* (Durham, NC: Duke University Press, 2001).

6. Spigel, *Welcome to the Dreamhouse*, 385.

7. Nick Srnicek, *Platform Capitalism* (Cambridge, UK: Polity, 2016); van Doorn, "Platform Labor."

8. Debbie Truong, "Students at a Virginia School Set out to Fight Segregation: Would They Succeed?," *Washington Post*, July 26, 2019.

9. "2013–2017 Demographic and Housing Estimates," American Community Survey, 2017 (US Census Bureau, n.d.); "2013–2017 Financial Characteristics ACS 5-Year Estimates," American Community Survey, 2017 (US Census Bureau, n.d.), compiled by Heather Suzanne Woods, using American Fact Finder.

10. "2013–2017 Demographic and Housing Estimates," American Community Survey, 2017 (US Census Bureau, n.d.); "2013–2017 Financial Characteristics ACS 5-Year Estimates," American Community Survey, 2017 (US Census Bureau, n.d.). Compiled by Heather Suzanne Woods, using American Fact Finder.

11. "2013–2017 Demographic and Housing Estimates," American Community Survey, 2017 (US Census Bureau, n.d.); "2013–2017 Financial Characteristics ACS 5-Year Estimates," American Community Survey, 2017 (US Census Bureau, n.d.), compiled by Heather Suzanne Woods, using American Fact Finder.

12. Truong, "Students at a Virginia School."

13. Rob Kitchin and Martin Dodge, *Code/Space: Software and Everyday Life* (Cambridge: MIT Press, 2011).

14. Brian Lonsway, *Making Leisure Work: Architecture and the Experience Economy* (London: Routledge, 2009), 5.

15. H. Ward Jandl, "Introduction," in *Yesterday's Houses of Tomorrow: Innovative American Homes, 1850 to 1950* (Washington, DC: Preservation Press, 1991).

16. From promotional literature for the Century 21 Plywood Home of Living Light, as quoted in Jandl 1991, *Yesterday's Houses of Tomorrow*, 207.

17. "Amazon Prime Air," Amazon, accessed June 26, 2019.

18. Nidhi Subbaraman, "'What If an Amazon Delivery Drone Crashes into Me?' and 6 Other Key Questions," *NBC News*, December 2, 2013.

19. Michael Sasso, "There Aren't Enough Drivers to Keep Up with Your Delivery Lifestyle," *Bloomberg*, June 13, 2018.

20. See, for instance, Nina Godlewski, "Amazon Employees Have Resorted to Urinating in Trash Cans in Some Warehouses," *Newsweek*, September 12, 2018; Hayley Peterson, "Amazon Delivery Drivers Say They Speed, Urinate in Bottles," *Business Insider*, September 12, 2018; Michael Sainato, "'We Are Not Robots': Amazon Warehouse Employees Push to Unionize," *The Guardian*, January 1, 2019, sec. Technology; Katie Schoolov, "How Amazon Is Fighting Back against Workers' Increasing Efforts to Unionize," CNBC, August 22, 2019.

21. Ceylan Yeginsu, "If Workers Slack Off, the Wristband Will Know (And Amazon Has a Patent for It)," *New York Times*, February 1, 2018.

22. Matt Novak, "Amazon Promised Drone Delivery in Five Years . . . Five Years Ago," *Gizmodo*, December 3, 2018.

23. Jeff Wilke, "A Drone Program Taking Flight," *About Amazon* (blog), June 5, 2019.

24. Nabyla Daidj, "Uberization (or Uberification) of the Economy," *Encyclopedia of Information Science and Technology* (IGI Global, 2018).

25. Ingrid Burrington, "Why Amazon's Data Centers Are Hidden in Spy Country," *The Atlantic*, January 8, 2016.

26. Yolande Strengers et al., "Protection, Productivity and Pleasure in the Smart Home: Emerging Expectations and Gendered Insights from Australian Early Adopters," *Proceedings of the 2019 CHI Conference on Human Factors in Computing Systems* (ACM, 2019), 645.

27. David Lyon, "Surveillance Technology and Surveillance Society," in *Modernity and Technology*, ed. Thomas J. Misa, Philip Brey, and Andrew Feenberg (Cambridge: MIT Press, 2003), 161–84.

28. John Edward Deukmedjian, "Making Sense of Neoliberal Securitization in Urban Policing and Surveillance," *Canadian Review of Sociology* 50, no. 1 (February 2013): 52–73.

29. George F. (Guy) McHendry, "Thank You for Participating in Security: Engaging Airport Security Checkpoints via Participatory Critical Rhetoric," *Cultural Studies ↔ Critical Methodologies* 16, no. 6 (December 2016): 9.

30. Nick Wingfield, "'Porch Pirates' Steal Holiday Packages as They Pile Up at Homes," *New York Times*, December 20, 2017.

31. Wingfield, "Porch Pirates."

32. Jennifer Medina, "Website Meant to Connect Neighbors Hears Complaints of Racial Profiling," *New York Times*, May 18, 2016; Jay McGregor, "Here's How Amazon's Ring Doorbell Police Partnership Affects You," *Forbes*, August 6, 2019.

33. McGregor, "Here's How Amazon's Ring."

34. Janis L. Edwards and Carol K. Winkler note that an ideograph is a concept, idea, or visual text that "develops through its usages and applications, operating as an abstraction and a fragment within the larger rhetorical environment." Importantly, like "ideology," "ideograph" is a rhetorical term for words, objects, and artifacts that appear to be both settled meaning while also being meaningfully vague—everyone "knows" what an ideograph is to the point its meaning is common sense, but what the ideograph means is also subject to change. "Representative Form and the Visual Ideograph: The Iwo Jima Image in Editorial Cartoons," *Quarterly Journal of Speech* 83 (1997): 297.

35. Katie Lambright, "Digital Redlining: The Nextdoor App and the Neighborhood of Make-Believe," *Cultural Critique* 103 (2019): 86.

36. van Doorn, "Platform Labor," 899.

37. Setha M. Low, "The Edge and the Center: Gated Communities and the Discourse of Urban Fear," *American Anthropologist* 103, no. 1 (2001): 55.

38. Shenle Pan, Vaggelis Giannikas, Yufei Han, Etta Grover-Silva, and Bin Qiao, "Using Customer-Related Data to Enhance e-Grocery Home Delivery," *Industrial Management & Data Systems* 117, no. 9 (2017): 1918.

39. Pan et al., "Using Customer-Related Data," 1918.

40. For problems with unattended delivery solutions, see Pan et al., "Using Customer-Related Data," 1918–1920.

41. Strengers et al., "Protection, Productivity, and Pleasure."

42. van Doorn, "Platform Labor."

43. van Doorn, "Platform Labor," 900.

44. van Doorn, "Platform Labor," 901.

45. van Doorn, "Platform Labor," 902.

46. van Doorn, "Platform Labor," 904.

47. van Doorn, "Platform Labor," 904.

48. Low, "The Edge and the Center."

49. Sophia Maalsen and Robyn Dowling, "Covid-19 and the Accelerating Smart Home," *Big Data & Society* 7, no. 2 (July 2020): 1–5.

50. Aaron van Dorn, Rebecca E. Cooney, and Miriam L. Sabin, "COVID-19 Exacerbating Inequalities in the US," *The Lancet* 395, no. 10232 (2020): 1243–44.

51. Annie Palmer, "How Amazon Managed the Coronavirus Crisis and Came Out Stronger," CNBS, September 29, 2020.

52. Edward Moyer, "Amazon Using Thermal Cameras for COVID-19 Temperature Checks," CNET, April 18, 2020.

53. Samuel Stebbins and Grant Suneson, "Jeff Bezos, Elon Musk among US Billionaires Getting Richer during Coronavirus Pandemic," *USA TODAY*, December 1, 2020, accessed January 25, 2021.

54. David Streitfeld, "How Amazon Crushes Unions," *New York Times*, March 16, 2021, sec. Technology.

55. Moyer, "Amazon Using Thermal Cameras."

56. Maalsen and Dowling, "Covid-19 and the Accelerating Smart Home," 4.

57. Kitchin & Dodge, *Code/Space*, 179.

Chapter 2

1. Armond R. Towns, "The 'Lumpenproletariat's Redemption': Black Radical Potentiality and LA Gang Tours," *Souls* 19, no. 1 (2017): 39–58; Pezzulo, "Touring 'Cancer Alley,'" *Text and Performance Quarterly* 23, no. 3 (2003): 228.

2. Pezzulo, "Touring 'Cancer Alley,'" 228.

3. Shannon Mattern, "A City Is Not a Computer," *Places*, February 2017.

4. John M. Carroll and Mary Beth Rosson, "Wild at Home: The Neighborhood as a Living Laboratory for HCI," *ACM Transactions on Computer-Human Interaction (TOCHI)* 20, no. 3 (2013): 1–28.

5. Carina Veeckman and Shenja van der Graaf, "The City as Living Laboratory: Empowering Citizens with the Citadel Toolkit," *Technology Innovation Management Review* 5 (March 23, 2015): 13–14.

6. Veeckman and van Der Graaf, "The City as Living Laboratory," 13–14.

7. Giorgia Nesti, "Co-Production for Innovation: The Urban Living Lab Experience," *Policy and Society* 37, no. 3 (2018): 310–25.

8. Giorgia Nesti, "Co-production for Innovation"; Carroll and Rosson, "Wild at Home," 2.

9. Caroline Gottschalk Drushke, "Agonistic Methodology: A Rhetorical Case Study in Agricultural Stewardship," in *Field Rhetoric: Ethnography, Ecology, & Engagement in the Places of Persuasion*, ed. Candice Rai and Caroline Gottschalk Druschke, 22–42 (Tuscaloosa: University of Alabama Press, 2018), 22.

10. Bill Hughes, "Being Disabled: Towards a Critical Social Ontology for Disability Studies," *Disability & Society* 22, no. 7 (December 2007): 674.

11. Carroll and Rosson, "Wild at Home."

12. Italics my own. "Imagine-SmartHomes," Imagine! Colorado, accessed March 12, 2020.

13. Italics my own. David J. Neuman, *Building Type Basics for College and University Facilities* (John Wiley & Sons, 2013), 243.

14. Duke Smart Home Program, "Visit the Duke Smart Home," Duke University Pratt School of Engineering, modified 2011–2020.

15. Veeckman and van der Graaf, "City as Living Laboratory," 9.

16. Sander van der Leeuw et al., "How Much Time Do We Have? Urgency and Rhetoric in Sustainability Science," *Sustainability Science* 7, no. 1 (2012): 115.

17. van der Leeuw et al., "How Much Time Do We Have?," 118.

18. Duke University, "Welcome to the Duke Smart Home: The Intersection of Technology and Sustainability at Duke University," Duke Smart Home, last edited 2019.

19. Duke Smart Home Program, "Live in the Smart Home," Duke University Pratt School of Engineering, modified 2011–2020.

20. William Hyung, "Smart Home Celebrates 2nd Birthday," *The Chronicle*, November 13, 2009.

21. Larry Mays, George Antoniou, and Andreas Angelakis, "History of Water Cisterns: Legacies and Lessons," *Water* 5, no. 4 (November 21, 2013): 1916–40.

22. Nesti, "Co-production for Innovation," 314.

23. Esteve Almirall and Jonathan Wareham, "Living Labs: Arbiters of Mid- and Ground-Level Innovation," *Technology Analysis & Strategic Management* 23, no. 1 (2011): 100.

24. Almirall and Wareham, "Living Labs," 88.

25. Ingrid Snijders, Lisette Wijnia, Remy M. J. P. Rikers, and Sofie M. M. Loyens, "Alumni Loyalty Drivers in Higher Education," *Social Psychology of Education* 22, no. 3 (2019): 607.

26. Hyung, "Smart Home Celebrates 2nd Birthday."

27. "HVAC Installation & Replacement," Home Depot Product Authority, LLC, last edited 2020.

28. Rachel Sereix, "Duke Smart Home Gears Up for a Year of Innovative Energy Storage Projects," *The Chronicle*, June 15, 2016.

29. danah boyd, *It's Complicated: The Social Lives of Networked Teens* (New Haven: Yale University Press, 2014); Alice E. Marwick and danah boyd, "Networked Privacy: How Teenagers Negotiate Context in Social Media," *New Media & Society* 16, no. 7 (2014): 1051–67.

30. Bruce Watson, "The Troubling Evolution of Corporate Greenwashing," *Chain Reaction*, no. 129 (April 2017): 38–40.

31. Michael Barbaro, "Home Depot to Go Green," *New York Times*, April 17, 2007.

32. "Siemens under Pressure Again over Australia Mega Coal Mine," *Reuters*, March 12, 2021.

33. Abbie Bennett, "NC Banned a Study on a Sea-Level Rise: Could It Mean More Hurricane Destruction?," *Raleigh News & Observer*, September 12, 2018.

34. North Carolina Department of Commerce, "Visitor Spending in North Carolina Grows at Near Record Pace," press release, May 7, 2019.

35. Kendall McGee, "2018 was a Record Year for NC Beaches, New Tourism Numbers Confirm," *WECT 6 News*, August 29, 2019.

36. Ben Finley, "AP Explains: Why North Carolina Is Vulnerable to Hurricanes," *AP News*, September 11, 2018.

37. Coral Davenport and Mark Landler, "Trump Administration Hardens Its Attack on Climate Science," *New York Times*, May 27, 2019.

38. Nesti, "Co-production for Innovation," 314.

39. "Community Centered Boards: What Is a Community Centered Board?," Colorado Department of Health Care Policy & Financing, accessed March 12, 2019.

40. Robert Leja, "Imagine! Colorado—Helping People toward a Life of New Possibilities," KUNC, September 2, 2014.

41. "Imagine! 2018 Annual Report," Imagine! Colorado, 2018.

42. "Imagine! 2018 Annual Report."

43. "The 2014 Stan Black Award," Community Foundation Boulder County, last updated 2020.

44. M. L. Harrison and Cathy Davis, *Housing, Social Policy, and Difference: Disability, Ethnicity, Gender, and Housing* (Bristol, UK: Policy Press, 2001), 117, 119.

45. Rob Imrie, "Housing Quality, Disability and Domesticity," *Housing Studies* 19, no. 5 (2004), 688.

46. Imrie, "Housing Quality," 689. Scholars note that the social process of constructing disability in a pejorative way is especially "curious" because "impairment is intrinsic to the human condition," such that "all human beings are likely . . . to experience bodily or physiological changes that, potentially, will affect the functioning of their bodies." Even absent a justice- and inclusivity-minded approach, there is a business case to be made for housing that supports all types of abilities.

47. Harrison and Davis, *Housing, Social Policy, and Difference*, 113.

48. Harrison and Davis, *Housing, Social Policy, and Difference*, 115.

49. Elizabeth E. Guffey, *Designing Disability: Symbols, Space, and Society* (London: Bloomsbury, 2018), 4.

50. Dan Goodley, "Dis/Entangling Critical Disability Studies," in *Culture—Theory—Disability: Encounters between Disability Studies and Cultural Studies*, ed. Hanjo Berressem, Moritz Ingwersen, and Anne Waldschmidt (Bielefeld, Germany: Transcript, 2017), 82.

51. Michael Rembis, "Introduction," in *Disabling Domesticity*, ed. Michael Rembis (New York: Palgrave Macmillan US, 2016), 6.

52. Rembis, "Introduction," 6.

53. Rembis, "Introduction," 6.

54. Andrew Marcum, "Rethinking the American Dream Home: The Disability Rights Movement and the Cultural Politics of Accessible Housing in the United States," in *Disabling Domesticity*, ed. Michael A. Rembis (New York: Palgrave Macmillan, 2016), 104.

55. Rembis, "Introduction," 6.

56. Emily Brooks, "'Don't Be a Knucklehead': Moralizing Disability in New Jersey's Pandemic Response and Rhetoric," *Disability Studies Quarterly* 41, no. 3 (2021).

57. Jennifer Kretchmer, Twitter post, December 18, 2019.

58. Marcum, "Rethinking the American Dream Home," 115.

59. Marcum, "Rethinking the American Dream Home," 103–35.

60. Imagine-SmartHomes, "Imagine!," accessed March 12, 2020.

61. I am borrowing the iterative, generative concept of poetic worldmaking by Michael Warner. Michael Warner, "Publics and Counterpublics," *Public Culture* 14, no. 1 (2002): 82.

62. Alison Kafer, "Crip Kin, Manifesting," *Catalyst: Feminism, Theory, Technoscience* 5, no. 1 (2019): 6.

63. Laura Mauldin, "A Feminist Technoscientific Approach to Disability and Caregiving in the Family," in *Disabling Domesticity*, ed. Michael Rembis (New York: Palgrave Macmillan US, 2016), 143.

64. Mauldin, "Feminist Technoscientific Approach," 143–44.

65. Sophia Maalsen, "Revising the Smart Home as Assemblage," *Housing Studies* (2019): 7.

66. Carroll and Rosson, "Wild at Home," 2.

67. Damien Smith Pfister, "Technoliberal Rhetoric, Civic Attention, and Common Sensation in Sergey Brin's 'Why Google Glass?,'" *Quarterly Journal of Speech* 105, no. 2 (2019): 182–203.

68. It is important to note here that the smart homes explored in this chapter are unique from other types of smart homes in this book, because they are not-for-profit. Nonprofits are increasingly asked to do more with less. Paul C. Light notes, in *Making Nonprofits Work*, that "the nonprofit sector has never been under greater pressure to improve," this "despite two decades of phenomenal growth." Shrinking state and federal funding paired with an increasingly competitive philanthropic landscape prompt many nonprofits to "innovate." Some nonprofits turn toward new technological advances to solve financial and logistical problems. As the collective social safety net shrinks, nonprofits must seek alternate support modes Both the HDSH and ICSH, for instance, rely upon external funding from stakeholders beyond the smart home. The ability to provide both essential services and radical experimentation is therefore dependent upon the generosity of external partners. These living labs must also serve their residents and community members within complex political processes, some of which directly contradict their stated missions. Smart homes are forced to innovate within the private sphere while simultaneously developing a compelling discursive strategy to expand the reach of the living laboratory.

Chapter 3

1. J. D. Biersdorfer, "Put Alexa and Siri to Work," *New York Times*, January 22, 2020.

2. Sarah Leavitt, *From Catharine Beecher to Martha Stewart: A Cultural History of Domestic Advice* (Chapel Hill: University of North Carolina Press, 2002), 8–10.

3. "Voice Assistants Used by 46% of Americans, Mostly on Smartphones," *Pew Research Center* (blog), accessed January 28, 2021.

4. Stacey Higginbotham, "How the Smart Home Has Changed in COVID Times," *Stacey on IoT* (blog), March 9, 2021.

5. Mishaela Robison, "Voice Assistants Have a Gender Bias Problem: What Can We Do about It?," *Brookings* (blog), December 9, 2020.

6. "Siri Does More than Ever. Even Before You Ask," Apple, accessed February 23, 2022.

7. "Alexa Features," Amazon, accessed February 23, 2022.

8. Muhammad Ashfaq, Jiang Yun, and Shubin Yu, "My Smart Speaker Is Cool! Perceived Coolness, Perceived Values, and Users' Attitude toward Smart Speakers," *International Journal of Human–Computer Interaction* 37, no. 6 (April 3, 2021): 560–73.

9. Heather Suzanne Woods, "Asking More of Siri and Alexa: Feminine Persona in Service of Surveillance Capitalism," *Critical Studies in Media Communication* 35, no. 4 (2018): 334–49.

10. Amy Schiller and John McMahon, "Alexa, Alert Me When the Revolution Comes: Gender, Affect, and Labor in the Age of Home-Based Artificial Intelligence," *New Political Science* 41, no. 2 (2019): 174.

11. Daniel M. Sutko, "Theorizing Femininity in Artificial Intelligence: A Framework for Undoing Technology's Gender Troubles," *Cultural Studies* 34, no. 4 (July 3, 2020): 583–84.

12. Robison, "Voice Assistants Have a Gender Bias Problem."

13. Yolande Strengers and Jenny Kennedy, *The Smart Wife: Why Siri, Alexa, and Other Smart Home Devices Need a Feminist Reboot* (Cambridge: MIT Press, 2020), 5.

14. As Schiller and McMahon suggest, "the introduction of artificial intelligence as *the* labor-saving device points to the increasingly computational and managerial nature of the domestic realm." Schiller and McMahon, "Alexa, Alert Me," 184.

15. Kylie Jarrett, *Feminism, Labour and Digital Media: The Digital Housewife* (London: Routledge), 2015.

16. Strengers and Kennedy, *Smart Wife.*

17. Jessica Enoch, *Domestic Occupations: Spatial Rhetorics and Women's Work*, Studies in Rhetorics and Feminism (Carbondale: Southern Illinois University Press, 2019).

18. Woods, "Asking More of Siri and Alexa."

19. Gina Masullo Chen, "Don't Call Me That: A Techno-Feminist Critique of the Term Mommy Blogger," *Mass Communication and Society* 16, no. 4 (2013): 510–32.

20. Hilary Bergen, "'I'd Blush If I Could': Digital Assistants, Disembodied Cyborgs and the Problem of Gender," *Word and Text: A Journal of Literary Studies and Linguistics* 6 (December 2016): 100.

21. Schiller and McMahon, "Alexa, Alert Me," 184; Strengers and Kennedy, *Smart Wife*; Sutko, "Theorizing Femininity in Artificial Intelligence," 575.

22. Strengers and Kennedy, *Smart Wife*, 17.

23. Sarah Sharma, "A Manifesto for the Broken Machine," *Camera Obscura: Feminism, Culture, and Media Studies* 35, no. 2 (104) (September 1, 2020): 175.

24. Sutko, "Theorizing Femininity in Artificial Intelligence," 577.

25. Thao Phan, "Amazon Echo and the Aesthetics of Whiteness," *Catalyst: Feminism, Theory, Technoscience* 5, no. 1 (April 1, 2019): 1–38; Taylor C. Moran, "Racial Technological Bias and the White, Feminine Voice of AI VAs," *Communication and Critical/Cultural Studies*, October 2, 2020, 1–18.

26. Sutko, "Theorizing Femininity in Artificial Intelligence," 568.

27. Nóra Ni Loideain and Rachel Adams, "From Alexa to Siri and the GDPR: The Gendering of Virtual Personal Assistants and the Role of Data Protection Impact Assessments," *Computer Law & Security Review* 36, no. 105366 (April 2020): 1–14; emphasis in original.

28. Strengers and Kennedy, *Smart Wife*, 2.

29. Nancy Folbre, "The Unproductive Housewife: Her Evolution in Nineteenth-Century Economic Thought," *Signs* 16, no. 3 (1991): 463–84.

30. Folbre, "Unproductive Housewife."

31. Not coincidentally, this figuration of both domesticity and femininity assumes in advance the perspective of white femininity. As Mukasa Mubirumusoke has argued, in the Western canon, domesticity serves as a site of meditation regarding the formation and definition of self and its relationship to others. Although the domestic may be underconsidered in this literature, it is of great importance, because it orients the philosophical imaginary around difference (e.g., "being at home . . . when confronted . . . by strangers") (76). The domestic sphere is therefore a foundational site of Black alienation, in part because domestic relation assumes in advance the subjectivity and agency of whiteness as ontology, and because this position figures Blackness as external and abject.

32. Ruth Schwartz Cowan, *More Work for Mother: The Ironies of Household Technology from the Open Hearth to the Microwave*, 2nd ed. (Basic Books, 1985), 201–2.

33. Cowan, *More Work for Mother.*

34. Phan, "Amazon Echo and the Aesthetics of Whiteness."

35. Phan, "Amazon Echo and the Aesthetics of Whiteness."

36. Phan, "Amazon Echo and the Aesthetics of Whiteness."

37. Jarrett, *Feminism, Labour and Digital Media*.

38. Martha Alter Chen, "Recognizing Domestic Workers, Regulating Domestic Work: Conceptual Measurement, and Regulatory Challenges," *Canadian Journal of Women and the Law* 23, no. 1 (2011): 168.

39. Enoch, *Domestic Occupations*, 183.

40. Folbre observes that "the 'cult of domesticity' contributed to the emergence of a distinctively female culture"; this female culture was associated with morality and goodness. "Ironically," Fobre writes, "the moral elevation of the home was accompanied by the economic devaluation of the work performed there." Folbre, "Unproductive Housewife," 465.

41. A. W. Geiger, "Sharing Chores a Key to Good Marriage, Say Majority of Married Adults," Pew Research Center, November 30, 2016.

42. Chen, "Recognizing Domestic Workers."

43. Enoch, *Domestic Occupations*, 6.

44. Enoch, *Domestic Occupations*, 12.

45. Enoch, *Domestic Occupations*, 11.

46. Jennifer Bailey Woodard and Teresa Mastin, "Black Womanhood: Essence and Its Treatment of Stereotypical Images of Black Women," *Journal of Black Studies* 36, no. 2 (2005): 264–81.

47. Amy G. Richter, *At Home in Nineteenth-Century America: A Documentary History* (New York: NYU Press, 2015), 11.

48. Enoch, *Domestic Occupations*, 12.

49. Phan, "Amazon Echo and the Aesthetics of Whiteness."

50. Strengers and Kennedy, *Smart Wife*, 2, 4.

51. Strengers and Kennedy, *Smart Wife*, 5.

52. Justin Eckstein and Anna M. Young, "Wasted Rhetoric," *Communication and Critical/Cultural Studies* 15, no. 4 (2018): 286.

53. Donovan Conley and Justin Eckstein, "Introduction: Soiled," in *Cookery: Food Rhetorics and Social Production*, Rhetoric, Culture, and Social Critique (Tuscaloosa: University of Alabama Press, 2020), 1–20.

54. Alice E. Marwick, "Instafame: Luxury Selfies in the Attention Economy," *Public Culture* 27, no. 1 (2015): 138.

55. Marwick, "Instafame," 140.

56. Marwick, "Instafame," 141, 156–57.

57. Maureen Ryan, "*Entertaining* Fantasies: Lifestyle and Social Life in 1980s America," *Journal of Communication Inquiry* 39, no. 1 (2015): 83.

58. Ryan, "*Entertaining* Fantasies," 84.

59. Ryan, "*Entertaining* Fantasies," 89.

60. "Works with Yummly," Whirlpool, accessed January 28, 2021.

61. "Yummly: Personalized Recipe Recommendations and Search," accessed January 28, 2021.

62. Natt Garun, "Whirlpool's New Smart Oven Works with Alexa and Yummly to Help You Avoid Burning Down Your Kitchen," *The Verge*, January 8, 2018.

63. Amazon has since deleted this listing on its platform, but links to the $300 Amazon Smart Oven can be found on secondhand sites, including Refinery29. See "Amazon Smart oven."

64. Whitson Gordon, "How Smart Should Your Home Be?," *Real Simple*, July 9, 2020.

65. For a fascinating treatment of the Jetsons' robot maid, Rosie, read Strengers and Kennedy, *Smart Wife*.

66. Gordon, "How Smart Should Your Home Be?"

67. *Consumer Reports*: CR's Ultimate Smart Home Toolkit, January 2021; Daniel Wroclawski, "Your Ultimate Smart Home Toolkit," *Consumer Reports*, January 2021.

68. Strengers and Kennedy, *Smart Wife*, 39.

69. Davin Heckman, *A Small World: Smart Houses and the Dream of the Perfect Day* (Durham, NC: Duke University Press Books, 2008), 44.

70. Consumer Reports, 29.

71. Woods, "Asking More of Siri and Alexa."

72. Bergen, "I'd Blush If I Could," 101–2.

73. David Pierce, "How Apple Finally Made Siri Sound More Human," *WIRED*, September 7, 2017, cited in Woods, "Asking More of Siri and Alexa."

74. Pierce, "How Apple Finally Made."

75. "Realdoll—The World's Finest Love Realdoll," RealDoll, accessed April 11, 2022.

76. Strengers and Kennedy, *Smart Wife*, 110.

77. Strengers and Kennedy, *Smart Wife*, 110.

78. Ry Crist, "Behind the Scenes of a Sexbot Factory," CNET, accessed January 28, 2021.

79. Jeff Elder, "AI-Powered Sex Robots Are Selling Well during Lockdown—Which Worries Some Experts, Who Say That They Can Introduce Some Surprising Regulatory Problems," *Business Insider*, June 13, 2020.

80. Schiller and McMahon, "Alexa, Alert Me," 182.

81. Strengers and Kennedy, *Smart Wife*, 5.

82. Lucy Worsley, *If Walls Could Talk: An Intimate History of the Home* (New York: Bloomsbury, 2013), 12.

83. Worsley, *If Walls Could Talk*, 12.

84. Joan Faber McAlister, "Figural Materialism: Renovating Marriage through the American Family Home," *Southern Communication Journal* 76, no. 4 (2011): 279–304.

85. McAlister, "Figural Materialism," 289–91.

86. McAlister, "Figural Materialism," 289.

87. McAlister, "Figural Materialism," 294.

88. McAlister, "Figural Materialism," 295.

89. Hannah Denham and Jaay Greene, "Did You Say, 'Hey, Siri'? Apple and Amazon Curtail Human Review of Voice Recordings," *Washington Post*, August 2, 2019.

90. Nick Duffy, "Your iPhone Now Wants to Know When You Have Sex (But Doesn't Mind If It's Gay or Straight)," *Pink News*, September 17, 2015.

91. Ana Valens, "The Apple Watch Tracks Calories Burned during Sex: What Does That
Mean for Data Privacy?" *Daily Dot*, October 10, 2020.

92. Valens, "The Apple Watch Tracks."

93. Jeremy Glass, "I Used a Fitness Tracker to Track My Activity Level during Sex," *Men's Health*, April 18, 2018.

94. Nick Wilson, E. Jane MacDonald, Osman David Mansoor, and Jane Morgan, "In Bed with Siri and Google Assistant: A Comparison of Sexual Health Advice," *BMJ*, December 13, 2017, 1–7.

95. Wilson et al., "In Bed with Siri," 2.

96. Schiller and McMahon, "Alexa, Alert Me," 175.

97. Bergen, "'I'd Blush If I Could,'" 100.

98. Bergen, "'I'd Blush If I Could,'" 96.

99. Strengers and Kennedy, *Smart Wife*, 77.

100. Bergen, "'I'd Blush If I Could,'" 96.

101. Ieva Spudyte, "I'd Blush If I Could: Closing Gender Divides in Digital Skills through Education—UNESCO Digital Library" (UNESCO, 2019).

102. "Meet Q: The First Genderless Voice," Genderlessvoice.com, accessed January 28, 2021.

103. "Meet Q: The First Genderless Voice."

104. "Meet Q: The First Genderless Voice."

105. "Meet Q: The First Genderless Voice."

106. Amanda Nell Edgar, *Culturally Speaking: The Rhetoric of Voice and Identity in a Mediated Culture* (Columbus: Ohio State University Press, 2019), 6.

107. Edgar, *Culturally Speaking*, 6–7.

108. Edgar, *Culturally Speaking*, 6–11.

109. Sarah Sharma, "A Manifesto for the Broken Machine," *Camera Obscura: Feminism, Culture, and Media Studies* 35, no. 2 (104) (September 1, 2020): 176.

110. Sharma, "Manifesto for the Broken Machine," 176.

111. Strengers and Kennedy, *Smart Wife*, 77.

Chapter 4

1. Davin Heckman, *A Small World: Smart Houses and the Dream of the Perfect Day* (Durham, NC: Duke University Press, 2008), 19.

2. Amy Kaplan, "Manifest Domesticity," *American Literature* 70, no. 3 (1998): 581.

3. Lauren Berlant, *The Queen of America Goes to Washington City: Essays on Sex and Citizenship* (Durham, NC: Duke University Press Books, 1997), 5. As Berlant writes, "the intimate public sphere of the U.S. present tense renders citizenship as a condition of social membership produced by personal acts and values, especially acts originating in or directed toward the family sphere. . . . In the new nostalgia-based fantasy nation of the 'American way of life,' the residential enclave where 'the family' lives usurps the modernist promise of the culturally vital, multiethnic city; in the new, utopian America, mass-mediated political identifications can only be rooted in traditional notions of home, family, and community."

4. Murray Goulden, "'*Delete the Family*': Platform Families and the Colonisation of the Smart Home," *Information, Communication & Society* (2019): 10.

5. Goulden further argues that certain parts of the smart home infrastructure are largely agnostic "in use," but "in placing a particular set of devices at the heart of the family, the gendering of those devices becomes salient." Goulden, "*Delete the Family*," 13.

6. Goulden, "*Delete the Family*," 8.

7. Nick Srnicek, *Platform Capitalism*, (Cambridge, UK: Polity, 2017), 54.

8. M. L. Harrison and Cathy Davis, *Housing, Social Policy, and Difference: Disability, Ethnicity, Gender, and Housing* (Bristol, UK: Policy Press, 2001), 115.

9. Sadowski, *Too Smart*, 136.

10. Sadowski, *Too Smart*, 135.

11. For more on the methodological processes associated with attending to spaces

and embedding the critical rhetorical self within those spaces, please see Candice Rai and Caroline Gottschalk Druschke, "On Being There: An Introduction to Studying Rhetoric in the Field," in *Field Rhetoric: Ethnography, Ecology, and Engagement in the Places of Persuasion*, ed. Candice Rai and Caroline Gottschalk Druschke (Tuscaloosa: University of Alabama Press, 2018), 1–21.

12. Rob Kitchin and Martin Dodge, *Code/Space: Software and Everyday Life* (Cambridge: MIT Press, 2011), 17.

13. Jathan Sadowski, *Too Smart: How Digital Capitalism Is Extracting Data, Controlling Our Lives, and Taking Over the World* (Cambridge: MIT Press, 2020), 142.

14. Srnicek, *Platform Capitalism*, 43–47.

15. Damien Smith Pfister and Misti Yang, "Five Theses on Technoliberalism and the Networked Public Sphere," *Communication and the Public* 3, no. 3 (2018): 249.

16. Pfister and Yang, "Five Theses," 254.

17. Pfister and Yang, "Five Theses."

18. Berlant, *Queen of America*, 10.

19. Berlant, *Queen of America*, 14.

20. Berlant, *Queen of America*, 14.

21. Sadowski, *Too Smart*, 135.

22. Jenny Odell, *How to Do Nothing: Resisting the Attention Economy* (Brooklyn: Melville House, 2019), 30.

23. Colin Lecher, "How Low Prices Could Make for an Antitrust Case against Amazon," *The Verge*, May 13, 2019.

24. Srnicek, *Platform Capitalism*, 46.

25. Carine Marzin, "Plug and Pray? A Disability Perspective on AI, Automated Decision-Making and Emerging Tech," European Disability Forum, May 2019, 20.

26. Pfister and Yang, "Five Theses," 250.

27. Sadowski, *Too Smart*, 147.

28. Jessica Enoch, *Domestic Occupations: Spatial Rhetorics and Women's Work*, Studies in Rhetorics and Feminisms (Carbondale: Southern Illinois University Press, 2019), 175.

29. Michele Lancione, "Radical Housing: On the Politics of Dwelling as Difference," *International Journal of Housing Policy* 20, no. 2 (2020): 275.

30. Valerie Padilla Carroll, "The Radical Possibilities of New (Feminist, Environmentalist) Domesticity: Housewifery as an Altermodernity Project," *Interdisciplinary Studies in Literature and Environment* 23, no. 1 (2016): 65.

31. Padilla Carroll, "Radical Possibilities," 52.

32. Padilla Carroll, "Radical Possibilities," 51–52.

33. Padilla Carroll, "Radical Possibilities," 60.

34. Padilla Carroll, "Radical Possibilities," 54.

35. bell hooks, *Yearning: Race, Gender, and Cultural Politics* (Boston: South End Press, 1999).

36. hooks, *Yearning*, 42.

37. Matthew Zook and Mark Graham, "Hacking Code/Space: Confounding the Code of Global Capitalism," *Transactions of the Institute of British Geographers* 43, no. 3 (2018): 391.

38. Zook and Graham, "Hacking Code/Space," 394.

39. Zook and Graham, "Hacking Code/Space," 394.

40. Ananya Roy, "Housing Justice: Towards a Field of Inquiry," in *Housing Justice in*

Unequal Cities, ed. Ananya Roy and Hilary Malson (Los Angeles: University of California, Institute on Inequality and Democracy, 2019), 14.

41. Roy, "Housing Justice," 16.

42. Roy, "Housing Justice," 14–15.

43. Sophia Maalsen, "Smart Housing: The Political and Market Responses of the Intersections between Housing, New Sharing Economies and Smart Cities," *Cities* 84 (2019): 1–7.

44. Maalsen, "Smart Housing," 4.

45. Adam Serwer, "The Amazon Union Exposes the Emptiness of 'Woke Capital' by Adam Serwer," *The Atlantic*, April 14, 2022.

46. Serwer, "Amazon Union."

47. Hilary Bergen, "'I'd Blush If I Could': Digital Assistants, Disembodied Cyborgs and the Problem of Gender," *Word and Text* 6 (2016): 106.

48. Bergen, "I'd Blush If I Could," 105.

49. Ben Tarnoff and Moira Weigel, "Opinion: Silicon Valley Workers Have Had Enough," *New York Times*, January 26, 2021.

50. bell hooks, *Belonging: A Culture of Place* (New York: Routledge, 2008), 29.

Bibliography

"2013–2017 Demographic and Housing Estimates." American Community Survey, 2017. US Census Bureau. Compiled by Heather Suzanne Woods, using American Fact Finder.

"2013–2017 Financial Characteristics ACS 5-Year Estimates." American Community Survey, 2017. US Census Bureau. Compiled by Heather Suzanne Woods, using American Fact Finder.

Almirall, Esteve, and Jonathan Wareham. "Living Labs: Arbiters of Mid- and Ground-Level Innovation." *Technology Analysis & Strategic Management* 23, no. 1 (2011): 87–102.

Amazon. "Amazon Prime Air." Accessed June 26, 2019.

Anand, Nikhil. "The Banality of Infrastructure." Items: Insights from the Social Sciences. Accessed June 26, 2019.

Andrejevic, Mark. "Foreword." In *Feminist Surveillance Studies*, edited by Rachel E. Dubrofsky and Shoshana Magnet. Durham, NC: Duke University Press, 2015.

Ashfaq, Muhammad, Jiang Yun, and Shubin Yu. "My Smart Speaker Is Cool! Perceived Coolness, Perceived Values, and Users' Attitude toward Smart Speakers." *International Journal of Human–Computer Interaction* 37, no. 6 (2021): 560–73.

Atha, Christine. "Dirt and Disorder: Taste and Anxiety in the Homes of the British Working Class." In *Atomic Dwelling: Anxiety, Domesticity, and Postwar Architecture*, edited by Robin Schuldenfrei, 207–26. New York: Routledge, 2012.

Austin, Patrick Lucas. "What Will Smart Homes Look like in 10 Years?" *Time*, July 25, 2019.

Baker, Paula. "The Domestication of Politics: Women and American Political Society, 1780–1920." *American Historical Review* 89, no. 3 (1984): 620–47.

Baydar, Gulsum. "Figures of Wo/Man in Contemporary Architectural Discourse." In *Negotiating Domesticity: Spatial Productions of Gender in Modern Architecture*, edited by Hilde Heynen and Gulsum Baydar, 30–47. London: Routledge, 2005.

Benfer, Emily, David Bloom Robinson, Stacy Butler, Lavar Edmonds, Sam Gilman, Katherine Lucas McKay, Zach Neumann, Lisa Owens, Neil Steinkamp, and Diane Yentel. "The COVID-19 Eviction Crisis: An Estimated 30–40 Million People in America Are at Risk." The Aspen Institute, August 7, 2020.

Bennett, Abbie. "NC Banned a Study on Sea-Level Rise: Could It Mean More Hurricane Destruction?" *Raleigh News & Observer*, September 12, 2018.

Bergen, Hilary. "'I'd Blush If I Could': Digital Assistants, Disembodied Cyborgs and the Problem of Gender." *Word and Text: A Journal of Literary Studies and Linguistics* 6 (December 2016): 95–113.

Berlant, Lauren. *The Queen of America Goes to Washington City: Essays on Sex and Citizenship*. Durham, NC: Duke University Press Books, 1997.

Berlant, Lauren, and Michael Warner. "Sex in Public." *Critical Inquiry* 24, no. 2 (1998): 547–66.

Berry, Mike, Mark Gibson, Anitra Nelson, and Ingrid Richardson. "How Smart Is 'Smart'?—Smart Homes and Sustainability." In *Steering Sustainability in an Urbanizing World: Policy, Practice and Performance*, edited by Anitra Nelson, 239–52. Hampshire, Burlington: Ashgate, 2007.

Biersdorfer, J. D. "Put Alexa and Siri to Work." *New York Times*, January 22, 2020.

Blair, Carole. "Contemporary U.S. Memorial Sites as Exemplars of Rhetoric's Materiality." In *Rhetorical Bodies*, edited by Jack Selzer and Sharon Crowley, 16–57. Madison: University of Wisconsin Press, 1999.

Bohn, Dieter. "Exclusive: Amazon Says 100 Million Alexa Devices Have Been Sold." The Verge, January 4, 2019.

Bowker, Geoffrey C., Karen Baker, Florence Millerand, and David Ribes. "Toward Information Infrastructure Studies: Ways of Knowing in a Networked Environment." In *International Handbook of Internet Research*, edited by Jeremy Hunsinger, Lisbeth Klastrup, and Matthew Allen, 97–117. Dordrecht: Springer Netherlands, 2010.

boyd, danah. *It's Complicated: The Social Lives of Networked Teens*. New Haven: Yale University Press, 2014.

Brock, André L. *Distributed Blackness: African American Cybercultures*. Critical Cultural Communication. New York: New York University Press, 2019.

Brooks, Emily. "'Don't Be a Knucklehead': Moralizing Disability in New Jersey's Pandemic Response and Rhetoric." *Disability Studies Quarterly* 41, no. 3 (2021).

Burrington, Ingrid. "Why Amazon's Data Centers Are Hidden in Spy Country." *The Atlantic*, January 8, 2016.

Carroll, John M., and Mary Beth Rosson. "Wild at Home: The Neighborhood as a Living Laboratory for HCI." *ACM Transactions on Computer-Human Interaction* 20, no. 3 (2013): 1–28.

CDC. "Interim Guidance on Unsheltered Homelessness and Coronavirus Disease 2019 (COVID-19) for Homeless Service Providers and Local Officials." Centers for Disease Control and Prevention, August 6, 2020.

CDC and HHS. "Temporary Halt in Residential Evictions to Prevent the Further Spread of COVID-19." Federal Register, September 4, 2020.

Chen, Gina Masullo. "Don't Call Me That: A Techno-Feminist Critique of the Term Mommy Blogger." *Mass Communication and Society* 16, no. 4 (2013): 510–32.

Chen, Martha Alter. "Recognizing Domestic Workers, Regulating Domestic Work: Conceptual Measurement, and Regulatory Challenges." *Canadian Journal of Women and the Law* 23, no. 1 (2011): 167–84.

CNBC Television. *Watch CNBC's Exclusive Interview with Amazon's SVP of Devices David Limp*. Accessed August 25, 2021.

"Community Centered Boards." Colorado Department of Health Care Policy & Financing. Accessed March 19, 2019.

Community Foundation Boulder County. "The 2014 Stan Black Award," 2014.

Conley, Donovan, and Justin Eckstein. "Introduction: Soiled." In *Cookery: Food Rhetorics and Social Production*, 1–20. Tuscaloosa: University of Alabama Press, 2020.

Couldry, Nick, and Ulises Ali Mejias. *The Costs of Connection: How Data Is Colonizing*

Human Life and Appropriating It for Capitalism. Stanford, CA: Stanford University Press, 2019.

Cowan, Ruth Schwartz. *More Work for Mother: The Ironies of Household Technology from the Open Hearth to the Microwave*. New York: Basic Books, 1985.

Crist, Ry. "Behind the Scenes of a Sexbot Factory." CNET. Accessed January 28, 2021.

Cuppini, Niccolò, and Mattia Frapporti. "Logistics Genealogies: A Dialogue with Stefano Harney." *Social Text* 36, no. 3 (September 1, 2018): 95–110.

Daidj, Nabyla. "Uberization (or Uberification) of the Economy." *Encyclopedia of Information Science and Technology* (Hershey, PA: IGI Global, 2018).

Darby, Sarah J. "Smart Technology in the Home: Time for More Clarity." *Building Research & Information* 46, no. 1 (2018): 140–47.

De Botton, Alain. *The Architecture of Happiness*. New York: Vintage Internat, 2008.

Denham, Hannah, and Jaay Greene. "Did You Say, 'Hey, Siri'? Apple and Amazon Curtail Human Review of Voice Recordings." *Washington Post*, August 2, 2019.

Deukmedjian, John Edward. "Making Sense of Neoliberal Securitization in Urban Policing and Surveillance." *Canadian Review of Sociology* 50, no. 1 (February 2013): 52–73.

Diaz, Von. "La Piña: A Pineapple Rum Inspired an Exquisite Daiquiri—and Reflections on History." *Garden & Gun*, August/September 2021.

Dickinson, Greg. *Suburban Dreams: Imagining and Building the Good Life*. Rhetoric, Culture, and Social Critique. Tuscaloosa: University of Alabama Press, 2015.

Doorn, Niels van. "Platform Labor: On the Gendered and Racialized Exploitation of Low-Income Service Work in the 'On-Demand' Economy." *Information, Communication & Society* 20, no. 6 (2017): 898–914.

Dorn, Aaron van, Rebecca E. Cooney, and Miriam L. Sabin. "COVID-19 Exacerbating Inequalities in the US." *The Lancet* 395, no. 10232 (2020): 1243–44.

Duke Smart Home Program. "Live in the Smart Home." Accessed January 13, 2021.

Duke Smart Home Program. "Visit the Duke Smart Home | Smart Home Program." Accessed January 13, 2021.

Eckstein, Justin, and Anna M. Young. "Wasted Rhetoric." *Communication and Critical/Cultural Studies* 15, no. 4 (2018): 274–91.

Edgar, Amanda Nell. *Culturally Speaking: The Rhetoric of Voice and Identity in a Mediated Culture*. Columbus: Ohio State University Press, 2019.

Edwards, Paul, Geoffrey Bowker, Steven Jackson, and Robin Williams. "Introduction: An Agenda for Infrastructure Studies." *Journal of the Association for Information Systems* 10, no. 5 (May 2009): 364–74.

Edwards, Paul N. "Infrastructure and Modernity: Force, Time, and Social Organization in the History of Sociotechnical Systems." In *Modernity and Technology*, edited by Thomas J. Misa, Philip Brey, and Andrew Feenberg, 185–225. Cambridge: MIT Press, 2003.

Elder, Jeff. "AI-Powered Sex Robots Are Selling Well during Lockdown—Which Worries Some Experts, Who Say That They Can Introduce Some Surprising Regulatory Problems." *Business Insider*, June 13, 2020.

Enoch, Jessica. *Domestic Occupations: Spatial Rhetorics and Women's Work*. Carbondale: Southern Illinois University Press, 2019.

Eubanks, Virginia. *Automating Inequality: How High-Tech Tools Profile, Police, and Punish the Poor*. New York: Picador St. Martin's Press, 2019.

Evans, James, Ross Jones, Andrew Karvonen, Lucy Millard, and Jana Wendler. "Living Labs and Co-Production: University Campuses as Platforms for Sustainability Science." *Current Opinion in Environmental Sustainability* 16 (October 2015): 1–6.

Feenberg, Andrew. "What Is Philosophy of Technology." Simon Fraser University, 2003.

Finley, Ben. "AP Explains: Why North Carolina Is Vulnerable to Hurricanes." *AP NEWS*, September 11, 2018.

Flores, Lisa A. "Creating Discursive Space through a Rhetoric of Difference: Chicana Feminists Craft a Homeland." *Quarterly Journal of Speech* 82, no. 2 (1996): 142–56.

Folbre, Nancy. "The Unproductive Housewife: Her Evolution in Nineteenth-Century Economic Thought." *Signs* 16, no. 3 (1991): 463–84.

Franklin, Seb. *Control: Digitality as Cultural Logic*. Cambridge: MIT Press, 2015.

Friedline, Terri, and Zibei Chen. "Digital Redlining and the Fintech Marketplace: Evidence from US Zip Codes." *Journal of Consumer Affairs* 55, no. 2 (2021): 366–88.

Garun, Natt. "Whirlpool's New Smart Oven Works with Alexa and Yummly to Help You Avoid Burning down Your Kitchen." The Verge, January 8, 2018.

Geiger, A. W. "Sharing Chores a Key to Good Marriage, Say Majority of Married Adults." Pew Research Center, November 30, 2016.

Glass, Jeremy. "I Used a Fitness Tracker to Track My Activity Level during Sex." *Men's Health*, April 18, 2018.

Godlewski, Nina. "Amazon Employees Have Resorted to Urinating in Trash Cans in Some Warehouses." *Newsweek*, September 12, 2018.

Goodley, Dan. "Dis/Entangling Critical Disability Studies." In *Culture—Theory—Disability: Encounters between Disability Studies and Cultural Studies*, edited by Hanjo Berressem, Moritz Ingwersen, and Anne Waldschmidt, 81–110. Bielefeld: Transcript, 2017.

Gordon, Whitson. "How Smart Should Your Home Be?" Real Simple, July 9, 2020.

Gottschalk Drushke, Caroline. "Agonistic Methodology: A Rhetorical Case Study in Agricultural Stewardship." In *Field Rhetoric: Ethnography, Ecology, and Engagement in the Places of Persuasion*, edited by Candice Rai and Caroline Gottschalk Druschke, 22–42. Tuscaloosa: University of Alabama Press, 2018.

Goulden, Murray. "'*Delete the Family*': Platform Families and the Colonisation of the Smart Home." *Information, Communication & Society* 24, no. 7 (September 26, 2019): 903–20.

Graham, Jefferson. "Smart Speakers: Is It Alexa, Google or Apple for You? Here's What You Need to Know to Decide." *USA TODAY*, November 24, 2020.

———. "You Just Got a Smart Speaker as a Holiday Present. Here's What You Need to Know." *USA TODAY*, December 25, 2020.

Gram-Hanssen, Kirsten, and Sarah J. Darby. "'Home Is Where the Smart Is'? Evaluating Smart Home Research and Approaches against the Concept of Home." *Energy Research & Social Science* 37 (March 2018): 94–101.

Halegoua, Germaine R. *The Digital City: Media and the Social Production of Place*. New York: NYU Press, 2020.

Harney, Stefano, and Fred Moten. *The Undercommons: Fugitive Planning & Black Study*. Wivenhoe, UK: Minor Compositions, 2013.

Harrison, M. L., and Cathy Davis. *Housing, Social Policy, and Difference: Disability, Ethnicity, Gender, and Housing*. Bristol, UK: Policy Press, 2001.

Haselton, Todd. "Now You Can Text Amazon Alexa from Your iPhone to Check the Weather, Buy Stuff and More—Here's How." *CNBC*, December 3, 2020.

Heckman, Davin. *A Small World: Smart Houses and the Dream of the Perfect Day*. Durham, NC: Duke University Press, 2008.

Heynen, Hilde. "Modernity and Domesticity: Tensions and Contradictions." In *Negotiating Domesticity: Spatial Productions of Gender in Modern Architecture*, edited by Hilde Heynen and Gulsum Baydar, 1–29. London: Routledge, 2005.

Higginbotham, Stacey. "How the Smart Home Has Changed in COVID Times." *Stacey on IoT* (blog), March 9, 2021.

Hoganson, Kristin L. *Consumers' Imperium: The Global Production of American Domesticity, 1865–1920*. Chapel Hill: University of North Carolina Press, 2007.

The Home Depot. "HVAC Installation and Replacement at The Home Depot," 2020.

hooks, bell. *Yearning: Race, Gender, and Cultural Politics*. Boston: South End Press, 1999.

Hyung, William. "Smart Home Celebrates 2nd Birthday." *The Chronicle*, November 13, 2009.

"Imagine! 2018 Annual Report." Annual Report. Imagine! Colorado, 2018.

Imbert, Fred. "Dow Drops Nearly 3,000 Points, as Coronavirus Collapse Continues; Worst Day since '87." CNBC, March 15, 2020.

Imrie, Rob. "Housing Quality, Disability and Domesticity." *Housing Studies* 19, no. 5 (2004): 685–90.

Ingraham, Christopher. "Analysis: World's Richest Men Added Billions to Their Fortunes Last Year as Others Struggled." *Washington Post*, January 1, 2021.

Isenstadt, Sandy. *The Modern American House: Spaciousness and Middle Class Identity*. Cambridge: Cambridge University Press, 2006.

Jandl, H. Ward. "Introduction." In *Yesterday's Houses of Tomorrow: Innovative American Homes, 1850 to 1950*. Washington, DC: Preservation Press, 1991.

———. *Yesterday's Houses of Tomorrow: Innovative American Homes, 1850 to 1950*. Preservation Press, 1991.

Jarrett, Kylie. *Feminism, Labour and Digital Media: The Digital Housewife*. New York: Routledge, 2015.

Jhally, Sut. *Media Education Foundation Transcript, Bell Hooks: Cultural Criticism & Transformation*, n.d.

Kafer, Alison. "Crip Kin, Manifesting." *Catalyst: Feminism, Theory, Technoscience* 5, no. 1 (2019): 1–37.

Kaplan, Amy. "Manifest Domesticity." *American Literature* 70, no. 3 (1998): 581–606.

Katuk, Norliza, Ku Ruhana Ku-Mahamud, Nur Haryani Zakaria, and Mohd Aizani Maarof. "Implementation and Recent Progress in Cloud-Based Smart Home Automation Systems." In *2018 IEEE Symposium on Computer Applications & Industrial Electronics (ISCAIE)*, Penang, Malaysia, 2018, 71–77.

Kennedy, Jenny, Bjorn Nansen, Michael Arnold, Rowan Wilken, and Martin Gibbs. "Digital Housekeepers and Domestic Expertise in the Networked Home." *Convergence: The International Journal of Research into New Media Technologies* 21, no. 4 (2015): 408–22.

Kitchin, Rob, and Martin Dodge. *Code/Space: Software and Everyday Life*. Cambridge: MIT Press, 2011.

Koetsier, John. "Alexa, Siri, Google Assistant: How the Top Smart Assistants Stack Up." *Forbes*, August 8, 2020.

Kretchmer, Jennifer. Twitter post. December 18, 2019.

Lambert, Lance. "Real Unemployment Rate Soars Past 20%—and the U.S. Has Now Lost 26.5 Million Jobs." *Fortune*, April 23, 2020.

Lancione, Michele. "Radical Housing: On the Politics of Dwelling as Difference." *International Journal of Housing Policy* 20, no. 2 (April 2, 2020): 273–89.

Leavitt, Sarah. *From Catharine Beecher to Martha Stewart: A Cultural History of Domestic Advice*. Chapel Hill: University of North Carolina Press, 2002.

Leeuw, Sander van der, Arnim Wiek, John Harlow, and James Buizer. "How Much Time Do We Have? Urgency and Rhetoric in Sustainability Science." *Sustainability Science* 7, no. 1 (2012): 115–20.

Leja, Robert. "Imagine! Colorado—Helping People toward a Life of New Possibilities." KUNC, September 2, 2014.

Levine, Helisse, and Anne G. Zahradnik. "Online Media, Market Orientation, and Financial Performance in Nonprofits." *Journal of Nonprofit & Public Sector Marketing* 24, no. 1 (2012): 26–42.

Light, Paul C. *Making Nonprofits Work: A Report on the Tides of Nonprofit Management Reform*. Washington, DC: Brookings Institution Press, 2011.

Locke, Taylor. "From Banana Bread to Lasagna: Top 10 Most-Googled Recipes while People Are Self-Quarantining." CNBC, April 26, 2020.

Loideain, Nóra Ni, and Rachel Adams. "From Alexa to Siri and the GDPR: The Gendering of Virtual Personal Assistants and the Role of Data Protection Impact Assessments." *Computer Law & Security Review* 36, no. 105366 (2020): 1–14.

Lonsway, Brian. *Making Leisure Work: Architecture and the Experience Economy*. London: Routledge, 2009.

Low, Setha M. "The Edge and the Center: Gated Communities and the Discourse of Urban Fear" *American Anthropologist* 103, no. 1 (2001): 45–58.

Luzum, Nathan, and Shagun Vashisth. "Duke University Admits Record-Low 5.7 Percent of Regular Decision Applicants." *The Chronicle*. Accessed January 13, 2021.

Lyon, David. "Surveillance Technology and Surveillance Society." In *Modernity and Technology*, edited by Thomas J. Misa, Philip Brey, and Andrew Feenberg, 161–84. Cambridge: MIT Press, 2003.

Maalsen, Sophia. "Revising the Smart Home as Assemblage." *Housing Studies* (2019): 1–16.

———. "Smart Housing: The Political and Market Responses of the Intersections between Housing, New Sharing Economies and Smart Cities." *Cities* 84 (2019): 1–7.

Maalsen, Sophia, and Robyn Dowling. "Covid-19 and the Accelerating Smart Home." *Big Data & Society* 7, no. 2 (2020): 1–5.

Marcum, Andrew. "Rethinking the American Dream Home: The Disability Rights Movement and the Cultural Politics of Accessible Housing in the United States." In *Disabling Domesticity*, edited by Michael A. Rembis, 103–35. New York: Palgrave Macmillan, 2016.

Marwick, Alice E. "Instafame: Luxury Selfies in the Attention Economy." *Public Culture* 27, no. 1 (2015): 137–60.

Marwick, Alice E., and danah boyd. "Networked Privacy: How Teenagers Negotiate Context in Social Media." *New Media & Society* 16, no. 7 (2014): 1051–67.

Marzin, Carine. "Plug and Pray? A Disability Perspective on AI, Automated Decision-Making and Emerging Tech." European Disability Forum, May 2019.

Mauldin, Laura. "A Feminist Technoscientific Approach to Disability and Caregiving in the Family." In *Disabling Domesticity*, edited by Michael A. Rembis, 139–61. New York: Palgrave Macmillan US, 2016.

McAlister, Joan Faber. "Figural Materialism: Renovating Marriage through the American Family Home." *Southern Communication Journal* 76, no. 4 (2011): 279–304.

McGee, Kendall. "2018 Was a Record Year for NC Beaches, New Tourism Numbers Confirm." *WECT News*, August 29, 2019.

McGregor, Jay. "Here's How Amazon's Ring Doorbell Police Partnership Affects You." Forbes, August 6, 2019.

McHendry, George F. (Guy). "Thank You for Participating in Security: Engaging Airport Security Checkpoints via Participatory Critical Rhetoric." *Cultural Studies ↔ Critical Methodologies* 16, no. 6 (December 2016): 548–59.

McKittrick, Katherine. "Mathematics Black Life." *Black Scholar* 44, no. 2 (Summer 2014): 16–28.

Medina, Jennifer. "Website Meant to Connect Neighbors Hears Complaints of Racial Profiling (Published 2016)." *New York Times*, May 18, 2016.

"Meet Q: The First Genderless Voice." Genderlessvoice.com. Accessed January 28, 2021.

Mitchell, Bruce, and Juan Franco. "HOLC 'Redlining' Maps: The Persistent Structure of Segregation and Economic Inequality." NCRC, March 2018.

Misa, Thomas J., Philip Brey, and Andrew Feenberg. *Modernity and Technology*. Cambridge: MIT Press, 2004.

Monahan, Torin. *Surveillance in the Time of Insecurity*. New Brunswick, NJ: Rutgers University Press, 2010.

Moran, Taylor C. "Racial Technological Bias and the White, Feminine Voice of AI VAs." *Communication and Critical/Cultural Studies*, October 2, 2020, 1–18.

Moyer, Edward. "Amazon Using Thermal Cameras for COVID-19 Temperature Checks." CNET, April 18, 2020.

"Nearly 20,000 Amazon Workers Tested Positive for COVID-19." *CBS News/AP*, October 2, 2020.

Nesti, Giorgia. "Co-Production for Innovation: The Urban Living Lab Experience." *Policy and Society* 37, no. 3 (2018): 310–25.

Neuman, David J. *Building Type Basics for College and University Facilities*. Hoboken, NJ: John Wiley & Sons, 2013.

North Carolina Department of Commerce. "Visitor Spending in North Carolina Grows at Near Record Pace." Press release, May 7, 2019.

Novak, Matt. "Amazon Promised Drone Delivery in Five Years . . . Five Years Ago." *Gizmodo*, December 3, 2018.

Nurock, Vanessa. "¿Puede prestar cuidados la inteligencia artificial?" *Cuadernos de relaciones laborales* 38, no. 2 (2020): 217–29.

Odell, Jenny. *How to Do Nothing: Resisting the Attention Economy*. Brooklyn: Melville House, 2019.

Olma, Sebastian. "Never Mind the Sharing Economy: Here's Platform Capitalism." Institute of Network Cultures, October 16, 2014.

Olmstead, Kenneth. "Voice Assistants Used by 46% of Americans, Mostly on Smartphones." Pew Research Center. Accessed January 28, 2021.

Padilla Carroll, Valerie. "The Radical Possibilities of New (Feminist, Environmentalist) Domesticity: Housewifery as an Altermodernity Project." *Interdisciplinary Studies in Literature and Environment* 23, no. 1 (2016): 51–70.

Palmer, Annie. "How Amazon Managed the Coronavirus Crisis and Came out Stronger." CNBS, September 29, 2020.

Pan, Shenle, Vaggelis Giannikas, Yufei Han, Etta Grover-Silva, and Bin Qiao. "Using Customer-Related Data to Enhance e-Grocery Home Delivery." *Industrial Management & Data Systems* 117, no. 9 (October 16, 2017): 1917–33.

Perez, Sarah. "Report: Voice Assistants in Use to Triple to 8 Billion by 2023." TechCrunch. Accessed January 28, 2021.

Petersen, Anne Helen. *Can't Even: How Millennials Became the Burnout Generation.* Boston: Houghton Mifflin Harcourt, 2020.

Peterson, Hayley. "Amazon Delivery Drivers Say They Speed, Urinate in Bottles." *Business Insider*, September 12, 2018.

Pezzullo, Phaedra C. "Touring 'Cancer Alley,' Louisiana: Performances of Community and Memory for Environmental Justice." *Text and Performance Quarterly* 23, no. 3 (2003): 226–52.

Pfister, Damien Smith. "Digitality, Rhetoric, and Protocological Fascism; Or, Fascist Ants & Democratic Cicadas." *Journal for the History of Rhetoric* 23, no. 1 (2020): 3–29.

———. "Technoliberal Rhetoric, Civic Attention, and Common Sensation in Sergey Brin's 'Why Google Glass?'" *Quarterly Journal of Speech* 105, no. 2 (2019): 182–203.

Pfister, Damien Smith, and Misti Yang. "Five Theses on Technoliberalism and the Networked Public Sphere." *Communication and the Public* 3, no. 3 (2018): 247–62.

Phan, Thao. "Amazon Echo and the Aesthetics of Whiteness." *Catalyst: Feminism, Theory, Technoscience* 5, no. 1 (2019).

Pierce, David. "How Apple Finally Made Siri Sound More Human." *WIRED*, September 7, 2017.

Rai, Candice, and Caroline Gottschalk Druschke. "On Being There: An Introduction to Studying Rhetoric in the Field." In *Field Rhetoric: Ethnography, Ecology, and Engagement in the Places of Persuasion*, edited by Candice Rai and Caroline Gottschalk Druschke, 1–21. Tuscaloosa: The University of Alabama Press, 2018.

Rai, Candice, and Caroline Gottschalk Druschke, eds. *Field Rhetoric: Ethnography, Ecology, and Engagement in the Places of Persuasion*. Tuscaloosa: University of Alabama Press, 2018.

"Realdoll—The World's Finest Love Realdoll." RealDoll. Accessed April 11, 2022.

Rembis, Michael. "Introduction." In *Disabling Domesticity*, edited by Michael Rembis, 1–23. New York: Palgrave Macmillan US, 2016.

Rice, Jeff. *Digital Detroit: Rhetoric and Space in the Age of the Network*. Carbondale: Southern Illinois University Press, 2012.

Richter, Amy G. *At Home in Nineteenth-Century America: A Documentary History.* New York: NYU Press, 2015.

Robison, Mishaela. "Voice Assistants Have a Gender Bias Problem: What Can We Do about It?" *Brookings* (blog), December 9, 2020.

Roy, Ananya. "Housing Justice: Towards a Field of Inquiry." In *Housing Justice in Unequal Cities*, edited by Ananya Roy and Hilary Malson, 13–19. Los Angeles: Institute on Inequality and Democracy, University of California, 2019.

Ryan, Maureen. "*Entertaining* Fantasies: Lifestyle and Social Life in 1980s America." *Journal of Communication Inquiry* 39, no. 1 (2015): 82–101.

Sadowski, Jathan. *Too Smart: How Digital Capitalism Is Extracting Data, Controlling Our Lives, and Taking over the World*. Cambridge: MIT Press, 2020.

Sadowski, Jathan, Yolande Strengers, and Jenny Kennedy. "More Work for Big Mother: Revaluing Care and Control in Smart Homes." *Environment and Planning A: Economy and Space*, June 14, 2021, 0308518X211022366.

Sainato, Michael. "'We Are Not Robots': Amazon Warehouse Employees Push to Unionize." *The Guardian*, January 1, 2019, sec. Technology.

Sasso, Michael. "There Aren't Enough Drivers to Keep Up with Your Delivery Lifestyle." *Bloomberg*, June 13, 2018.

Schiller, Amy, and John McMahon. "Alexa, Alert Me When the Revolution Comes: Gender, Affect, and Labor in the Age of Home-Based Artificial Intelligence." *New Political Science* 41, no. 2 (2019): 173–91.

School, Katie. "How Amazon Is Fighting Back against Workers' Increasing Efforts to Unionize." CNBC, August 22, 2019.

Sereix, Rachel. "Duke Smart Home Gears Up for Year of Innovative Energy Storage Projects." *The Chronicle*, June 15, 2016.

Sharma, Sarah. "A Manifesto for the Broken Machine." *Camera Obscura: Feminism, Culture, and Media Studies* 35, no. 2 (2020): 171–79.

Snijders, Ingrid, Lisette Wijnia, Remy M. J. P. Rikers, and Sofie M. M. Loyens. "Alumni Loyalty Drivers in Higher Education." *Social Psychology of Education* 22, no. 3 (2019): 607–27.

Spigel, Lynn. *Make Room for TV: Television and the Family Ideal in Postwar America*. Chicago: University of Chicago Press, 1992.

———. *Welcome to the Dreamhouse: Popular Media and Postwar Suburbs*. Console-Ing Passions: Television and Cultural Power. Durham, NC: Duke University Press, 2001.

Spillers, Hortense J. *Black, White, and in Color: Essays on American Literature and Culture*. Chicago: University of Chicago Press, 2003.

Spudyte, Ieva. "I'd Blush If I Could: Closing Gender Divides in Digital Skills through Education—UNESCO Digital Library." (UNESCO, 2019).

Srnicek, Nick. *Platform Capitalism*. 1st ed. Cambridge, UK: Polity, 2016.

Stebbins, Samuel, and Grant Suneson. "Jeff Bezos, Elon Musk among US Billionaires Getting Richer during Coronavirus Pandemic." *USA TODAY*. December 1, 2020. Accessed January 25, 2021.

Strengers, Yolande. *Smart Energy Technologies in Everyday Life: Smart Utopia?* Dordrecht: Springer, 2013.

Strengers, Yolande, and Jenny Kennedy. *The Smart Wife: Why Siri, Alexa, and Other Smart Home Devices Need a Feminist Reboot*. Cambridge: MIT Press, 2020.

Strengers, Yolande, Jenny Kennedy, Paula Arcari, Larissa Nicholls, and Melissa Gregg. "Protection, Productivity and Pleasure in the Smart Home: Emerging

Expectations and Gendered Insights from Australian Early Adopters." *Proceedings of the 2019 CHI Conference on Human Factors in Computing Systems* (2019): 1–13.

Subbaraman, Nidhi. "'What If an Amazon Delivery Drone Crashes into Me?' and 6 Other Key Questions." *NBC News*, December 2, 2013.

Sutko, Daniel M. "Theorizing Femininity in Artificial Intelligence: A Framework for Undoing Technology's Gender Troubles." *Cultural Studies* 34, no. 4 (2020): 567–92.

Szymkowski, Sean. "Lowest Gas Prices in the US Falls [*sic*] to Just 91 Cents a Gallon." *CNET Roadshow*. Accessed January 25, 2021.

Tapper, James. "Alexa, Siri . . . Elsa? Children Drive Boom in Smart Speakers." *The Guardian*, October 18, 2020.

Tarnoff, Ben, and Moira Weigel. "Opinion | Silicon Valley Workers Have Had Enough." *New York Times*, January 26, 2021.

Thompson, Derek. "Go Home Now." *The Atlantic*, December 1, 2020.

Towns, Armond R. "The 'Lumpenproletariat's Redemption': Black Radical Potentiality and LA Gang Tours." *Souls* 19, no. 1 (2017): 39–58.

Truong, Debbie. "Students at a Virginia School Set Out to Fight Segregation. Would They Succeed?" *Washington Post*, July 26, 2019.

"US Oil Futures Plunge below $0 for First Time in History." *Al Jazeera*, April 27, 2020.

Vakil, Keya. "'Those Numbers Take Your Breath Away': Why Black Americans Are Dying from COVID-19 at Alarming Rates." *COURIER*, April 20, 2020.

Valens, Ana. "The Apple Watch Tracks Calories Burned during Sex: What Does That Mean for Data Privacy?" *Daily Dot*, October 10, 2020.

Valentino-DeVries, Jennifer, Denise Lu, and Gabriel J. X. Dance. "Location Data Says It All: Staying at Home during Coronavirus Is a Luxury." *New York Times*, April 3, 2020.

Veeckman, Carina, and Shenja van der Graaf. "The City as Living Laboratory: Empowering Citizens with the Citadel Toolkit." *Technology Innovation Management Review* no. 5 (2015): 6–17.

"'Victory Gardens' for the War against COVID-19." *CBS News*, April 5, 2020.

Warner, Michael. "Publics and Counterpublics." *Public Culture* 14, no. 1 (2002): 49–90.

Watson, Bruce. "The Troubling Evolution of Corporate Greenwashing." *Chain Reaction*, no. 129 (April 2017): 38–40.

Wilke, Jeff. "A Drone Program Taking Flight." *About Amazon* (blog), June 5, 2019.

Wilson, Nick, E. Jane MacDonald, Osman David Mansoor, and Jane Morgan. "In Bed with Siri and Google Assistant: A Comparison of Sexual Health Advice." *BMJ*, December 13, 2017: 359.

Wingfield, Nick. "'Porch Pirates' Steal Holiday Packages as They Pile Up at Homes." *New York Times*, December 20, 2017.

Woods, Heather Suzanne. "Asking More of Siri and Alexa: Feminine Persona in Service of Surveillance Capitalism." *Critical Studies in Media Communication* 35, no. 4 (2018): 334–49.

"Works with Yummly." *Whirlpool*. Accessed January 28, 2021.

Worsley, Lucy. *If Walls Could Talk: An Intimate History of the Home*. New York: Bloomsbury USA, 2013.

Wroclawski, Daniel. "Your Ultimate Smart Home Toolkit." *Consumer Reports*, January 2021.

Yeginsu, Ceylan. "If Workers Slack Off, the Wristband Will Know (And Amazon Has a Patent for It)." *New York Times*, February 1, 2018.

"Yummly: Personalized Recipe Recommendations and Search." Yummly. Accessed January 28, 2021.

Zook, Matthew, and Mark Graham. "Hacking Code/Space: Confounding the Code of Global Capitalism." *Transactions of the Institute of British Geographers* 43, no. 3 (September 2018): 390–404.

Index